SPSS

FOR

DUMMIES®

2ND EDITION

by Arthur Griffith

WILEY

John Wiley & Sons, Inc.

SPSS For Dummies®, 2nd Edition

Published by
John Wiley & Sons, Inc.
111 River Street
Hoboken, NJ 07030-5774

www.wiley.com

Copyright © 2010 by John Wiley & Sons, Inc., Hoboken, New Jersey

Published by John Wiley & Sons, Inc., Hoboken, New Jersey

Published simultaneously in Canada

For general information on our other products and services, please contact our Customer Care Department within the U.S. at 877-762-2974, outside the U.S. at 317-572-3993, or fax 317-572-4002.

For technical support, please visit www.wiley.com/techsupport.

Wiley publishes in a variety of print and electronic formats and by print-on-demand. Some material included with standard print versions of this book may not be included in e-books or in print-on-demand. If this book refers to media such as a CD or DVD that is not included in the version you purchased, you may download this material at http://booksupport.wiley.com. For more information about Wiley products, visit www.wiley.com.

Library of Congress Control Number: 2009940867

ISBN 978-0-470-48764-8 (pbk); ISBN 978-0-470-59948-8 (ebk); ISBN 978-0-470-59997-6 (ebk); ISBN 978-0-470-59999-0 (ebk)

Manufactured in the United States of America

10 9 8 7 6 5

WILEY

About the Author

Arthur Griffith is a computer programmer and a writer. He is the author of twelve books and the coauthor of three. His education was many years ago in a land far away, and he has a degree in Computer Science and Mathematics.

During his years as a computer programmer, he developed systems as varied as nuclear power-plant construction accounting, missile guidance, remote control of cable-TV set-top boxes, and satellite communications control. All the work he did with computer programming required the use of mathematics and the ability to explain complex concepts in simple language.

He moved to Alaska in an attempt to retire, but that project failed. He is now developing online tutorials and writing books, like this one.

He and his wife, Mary, live high up on a ridge in remote Alaska, with moose and bear in the yard and eagles hunting from the roof.

Dedication

To the nurses of the South Peninsula hospital, who kept this Dummy alive and able to write a book.

Author's Acknowledgments

I need to thank the professionals at SPSS. In particular, Matthew Madden made information available that was invaluable in the writing of this book. Aaron Rangel operated an excellent program that provided access to a pre-release copy of the software. Sarah Tomashek connected me with the information necessary to write and made certain I was able to receive a copy of the software. Beth Narrish was kind enough to provide me with early access to a copy of the final release of SPSS.

Amy Fandrei at John Wiley first came up with the idea that this book should be written.

Christopher Morris at John Wiley was patient and understanding when delivery deadlines whisked by. He kept everything sane while I juggled the chapters of the book with the delivery schedule. Rhonda Smith checked the book for mistakes, so if you find anything wrong, it's her fault. (Kidding. Just kidding.)

Publisher's Acknowledgments

We're proud of this book; please send us your comments at http://dummies.custhelp.com. For other comments, please contact our Customer Care Department within the U.S. at 877-762-2974, outside the U.S. at 317-572-3993, or fax 317-572-4002.

Some of the people who helped bring this book to market include the following:

Acquisitions and Editorial

Sr. Project Editor: Christopher Morris

Acquisitions Editor: Amy Fandrei

Sr. Copy Editor: Barry Childs-Helton

Technical Editor: Rhonda Smith

Editorial Manager: Kevin Kirschner

Editorial Assistant: Amanda Graham

Sr. Editorial Assistant: Cherie Case

Cartoons: Rich Tennant
    (www.the5thwave.com)

Composition Services

Project Coordinator: Patrick Redmond

Layout and Graphics: Melissa K. Jester, Ronald G. Terry, Christine Williams

Proofreader: Christine Sabooni

Indexer: Broccoli Information Mgt.

Publishing and Editorial for Technology Dummies

   Richard Swadley, Vice President and Executive Group Publisher

   Andy Cummings, Vice President and Publisher

   Mary Bednarek, Executive Acquisitions Director

   Mary C. Corder, Editorial Director

Publishing for Consumer Dummies

   Kathleen Nebenhaus, Vice President and Executive Publisher

Composition Services

   Debbie Stailey, Director of Composition Services

Contents at a Glance

Table of Contents

Introduction

*G*ood news! You don't have to know diddly-squat about statistics to be able to come up with well-calculated conclusions and display them in fancy graphs. All you need is the IBM SPSS Statistics software and a bunch of numbers. This book shows you how to type the numbers, click options in the menus, and produce brilliant statistics. It really is as simple as that.

About This Book

This is fundamentally a reference book. Parts of the book are written as standalone tutorials to make it easy for you to get into whatever you're after. Once you're up and running with SPSS, you can skip around and read just the sections you need. You really don't want to read straight through the entire book. That way leads to boredom. I know — I went straight through everything to write the book, and believe me, you don't want to do that.

The book was designed to be used as follows:

1. Read the opening chapter so you'll understand what SPSS is. I tried to leave out the boring parts.

2. If SPSS is not already installed, you may need to read about installing it. That's Chapter 2.

3. Read the stuff in Chapter 4 about defining variables and entering data. It all makes sense once you get the hang of it, but the process seems kind of screwy until you see how it works.

4. Skip around to find the things you want to do.

I would mention that you could skip the introduction, but it's too late for that. Besides, you may find some information here that could be useful.

One thing that needs to be clear from the beginning: This book is not about statistics. You will not find one explanation of statistical theory or how calculations are performed. This book is about the things you can do to command SPSS software to calculate statistics for you. The inside truth is that you can be as dumb as a post about statistical calculation techniques and still use SPSS to produce some nifty stats. You have my permission to stop thinking right now.

However, if you decide to study the techniques of statistical calculation, you'll be able to understand what SPSS does to produce numbers. Your main advantage in understanding the process to that degree of detail is that you'll be able to choose a calculation method that more closely models the reality you are trying to analyze — if you're interested in reality, of course.

Shortly before this book went to press, SPSS Inc. was acquired by IBM Corporation. The name of the product referred to throughout this book was changed from *PASW Statistics* to *IBM SPSS Statistics*.

About the Data

Throughout the book you will find examples that use data stored in files. These files are freely available to you. Most of the files are installed with IBM SPSS Statistics in the SPSS installation directory which, by default, is `\Program Files\PASW` (unless you chose another location during installation). A few files were designed for this book and are available on the book's companion Web site.

In every case, the files were especially designed to demonstrate some specific capability of IBM SPSS Statistics.

If you have a question about the data, or if you wish to contact me about some other question you may have, you can reach me at the following e-mail address:

`arthur@belugalake.com`

Who This Book Is For

In general terms, this book is for anyone new to SPSS. No prior knowledge of statistics or mathematics is needed, or even expected. In specific terms, this book was written with two groups in mind: students who are not majoring in mathematics but are instructed to use SPSS, and office workers who are instructed to use SPSS to analyze some data.

For most people who generate statistics, the complexity of using the software becomes an obstacle. My purpose in writing this book is to show you how to move that obstacle out of the way with minimum effort.

How This Book Is Organized

This book was written so you could read the first part, to get yourself started with SPSS, and then jump around to the other parts as needed. SPSS is a huge piece of software, and you certainly don't want to try to use everything.

The book is filled with step-by-step procedures that you can follow to see how SPSS operates. After you use the provided sample data and step through an example, you'll have a handle on how to apply those steps to your data.

The parts of the book divide the information about SPSS into its major categories. The chapters in each part further divide the information into smaller categories.

Part 1: The Fundamental Mechanics of SPSS

The first part is the only one intended to be read straight through. You can gloss over the installation if you already *have* SPSS installed, but it's still worth familiarizing yourself with the configuration options. You'll come across these configurations later and will need to know what can be changed. This is the only place in the book where you find an obsessively complete example of using SPSS — starting with the entering of gathered data and ending with the generation of rudimentary analyses.

Part 11: Getting Data In and Out of SPSS

Input can be tricky. Variables are defined by type and size and a few other things. Part II shows you how to enter data through the main SPSS window or load it from a file. In fact, you can read data from several kinds of files. You can also write data to several kinds of files.

Part 111: Graphing Data

In Part III you see how to produce graphs. A large part of the job performed by SPSS is displaying data in graphic formats. SPSS can produce lots of different kinds of graphs. Fortunately, it's easy to do — you simply select variable names and specify how you want them displayed.

Part IV: Analysis

Hidden down inside SPSS are lots of statistical methods. I tell ya, this program manufactures numbers like McDonald's manufactures hamburgers. Part IV explains how to manufacture the numbers you want.

Part V: Programming SPSS with Command Syntax

Part V shows you how to use the SPSS internal command language. You can record procedures in Command Syntax and execute them at will. You can do anything with a Command Syntax program that you can do with the mouse and keyboard. And then some.

Part VI: Programming SPSS with Python and Scripts

Part VI is BASIC talk about programming and scripting SPSS. Anything you can do with Command Syntax or with the mouse and keyboard, you can also do in the Python programming language, and you can schedule the scripts to execute automatically under various circumstances. The scripting languages of SPSS are Sax BASIC and Python 2.6.

Part VII: The Part of Tens

If you're in quest of some new capabilities and resources for SPSS, check out Part VII. It's all about the add-ons for SPSS and the locations on the Internet where you can find useful stuff.

Icons Used in This Book

You should keep this information in mind. It's important to what you're doing.

Skip these unless the text makes you curious. This icon highlights unnecessarily geeky information, but I had to include it to complete the thought.

A tip highlights a point that can save you time and effort.

As is traditional with warnings, these offer information about something that can sneak up and bite you.

Where to Go from Here

Read the first chapter. Then, if necessary, install SPSS, referring to Chapter 2. Work through the example in Chapter 3.

Now you're up and running. Figure out what you want to do with SPSS — and then refer to the sections of the book that are necessary for doing it. For some tasks (such as programming in Python), you need to read an entire chapter. For other jobs, you need to read only a single section. This book's Cheat Sheet can be found at www.dummies.com/cheatsheet/spss.

Its companion Web site can be found at www.dummies.com/go/pasw.

Part I

The Fundamental Mechanics of SPSS

The 5th Wave   By Rich Tennant

"Our customer survey indicates 30% of our customers think our service is inconsistent, 40% would like a change in procedures, and 50% think it would be real cute if we all wore matching colored vests."

In this part . . .

This is a look at SPSS Statistics from 10,000 feet. Even if you know nothing whatsoever about SPSS, after you read this part you'll have a good idea of how it all works. You might not have all the niggling details, but you'll have the lowdown on the general operation of SPSS. Everything else you find out about SPSS will fit in the structure you build for yourself by reading Part I.

This is the only part of the book intended to be read straight through. The only optional subject in Part I is the description of the installation, which you won't have to bother with if you've already installed SPSS — and if you haven't yet, it'll help you get that done.

Chapter 1

Introducing IBM SPSS Statistics

* *

A statistic is a number. A raw statistic is a measurement of some sort. It is fundamentally a count of something — occurrences, speed, amount, or whatever. IBM SPSS Statistics is a piece of software that takes in raw data and combines them into new statistics that can be used as predictors.

"There are three kinds of lies: lies, damn lies, and statistics." That statement is often attributed to Mark Twain, but that's not quite right. Mark Twain did say it, but he attributed it to someone else. He indirectly attributed it to Disraeli, but his attribution was vague, and the original statement, if it exists, can't be located. Speaking statistically, the odds are we'll never know who said it first.

Garbage In, Garbage Out

Statistical analysis is like a sewer. What you get out of it depends on what you put into it.

Eighty-two percent of all statistics are made up on the spot to try to prove a point.

If you're not careful, you can conclude just about anything from your data and your calculations. SPSS performs calculations for you, but the raw data, and which calculations are performed, are up to you.

Let me show you a simple example of using raw data to produce an obviously wrong conclusion. Suppose you want to demonstrate, by sampling, that every odd number is prime. (A prime number can be evenly divided only by 1 and itself.) The first thing to do is gather a collection of data points, as shown in Table 1-1.

Table 1-1	Odd Numbers and Whether They Are Prime	
Number	**Prime?**	**Comment**
1	Yes	It fits the definition exactly
3	Yes	It is certainly both odd and prime
5	Yes	It fits the pattern of primes
7	Yes	So far, so good
9	No	Must be a bad data point, so throw it out
11	Yes	Now we're back on track
13	Yes	Looking good

Lots of things are already wrong with the data in Table 1-1. For one, the sample is too small. For another, the sampling cannot be considered random. All too often it happens that data points are omitted if they don't fit a preconceived conclusion. The result of the data in this table can be used as "proof" of a "fact" that is dead wrong.

This book is not about the accuracy, correctness, or completeness of the input data. Your data is up to you. This book shows you how to take the numbers you already have, put them into SPSS, crunch them, and display the results in a way that makes sense. Gathering valid data and figuring out which crunch to use is up to you.

Where Did SPSS Come From?

SPSS is probably older than you are. In 2009 it became 40 years old, and the average age of an American is 35.3.

At Stanford University in the late 1960s, Norman H. Nie, C. Hadlai (Tex) Hull, and Dale H. Bent developed the original software system named Statistical Package for the Social Sciences (SPSS). They needed to analyze a large volume of social science data, so they wrote software to do it. The software package caught on with other folks at universities and, consistent with the open-source tradition of the day, the software spread through universities around the country.

The three men produced a manual in the 1970s, and the software's popularity took off. A version of it existed for each of the different kinds of mainframe computers existing at the time. Its popularity spread from universities into other areas of government, and it began to leak out into private enterprise.

In the 1980s, a version of the software was moved to the personal computer. In 2008, the name was briefly changed to Predictive Analysis Software (PASW). In 2009, SPSS Inc. was acquired by IBM Corporation and the name of the product was returned to the more familiar SPSS. The official name of the software today is IBM SPSS Statistics.

Maybe it has been continuously successful because the software does such a good job of making predictions, and the SPSS people could always figure out what they should do next.

The practical application of the software has always been to attempt to predict the future. Predictive models are used on business data to identify both risks and opportunities. Relationships among many factors are analyzed to guide decision-makers in selecting from among a number of possible actions.

The software is available in several forms — single user, multiuser, client-server, student version, and so on. The software also has a number of special purpose add-ons available. You can find out about them all at the following Web site: www.spss.com

The Four Ways to Talk to SPSS

More than one way exists for you to command SPSS to do your bidding. You can use any of four approaches to perform any of the SPSS functions, but the one you should choose depends not only on which interface you prefer, but also (to an extent) on the task you want performed. The available interfaces are as follows:

- ✔ **GUI (graphic user interface):** SPSS has a windowing interface; you can issue commands by using the mouse to make menu selections that cause dialog boxes to appear. This is a fill-in-the-blanks approach to statistical analysis that guides you through the process of making choices and selecting values. The advantage of the GUI approach is that, at each step, SPSS makes sure you enter everything necessary before you can proceed to the next step. This is the preferred interface for those just starting out — and if you don't go into depth with SPSS, this may be the only interface you ever use.

- ✔ **Syntax:** This is the internal language used to command actions from SPSS. It is the command syntax of SPSS, hence its name. It's often referred to as *the command language*. You can use the Syntax command language to enter instructions into SPSS and have it do anything it's capable of doing. In fact, when you select from menus and dialog boxes to command SPSS, you're actually generating Syntax commands internally that do your bidding. That is, the GUI is nothing more than the *front* *end* of a Syntax command-writing utility. Writing (and saving) command-language programs is a good way to create processes that you expect to repeat. You can even grab a copy of the Syntax commands generated from the menu and save them to be repeated later.

✔ **Python:** This is a general-purpose language that has a collection of SPSS modules written for it; you can use it to write programs that work inside SPSS. You can also run Python with the Syntax language to command SPSS to perform statistical functions. One advantage of using Python is that it's a modern language, complete with the power and convenience that come with such languages, including the capability of constructing a more readable program. In addition, because Python is a general-purpose language, you can read and write data in other applications and in files.

✔ **Scripts:** The items that SPSS calls *scripts* are actually programs written in BASIC. This language is simple and many people are familiar with it. Also, a BASIC program can be written as an *autoscript* — a script that executes automatically whenever SPSS produces certain output. Both BASIC and Python are scripting languages, but where the SPSS documentation talks about a script, it is referring to a BASIC program.

What You Can and Cannot Do with SPSS

The full-blown SPSS package comes in many parts. The *Base system* is the center around which the rest of SPSS revolves. If you have SPSS, you have a Base system.

You may also have one or more add-ons. With only one exception — the Python programming language, which requires some additional software available for free on the SPSS distribution CD — everything described in this book is included in the Base system, so you will be able to do anything you read about. Chapter 20 describes other modules you can add to your Base system.

SPSS works with numbers. Only. If you cannot express your information as a number, you can't run it through SPSS. You will see names and descriptions seemingly being processed by SPSS, but that's because each name has been assigned a number. (Sneaky.) That's why survey questions are written like this: "How *much* do you enjoy eating rhubarb? Select your answer: Very much, sort of, don't care, not really, I hate the stuff." A number is assigned to each of the possible answers, and these numbers are fed through the statistical process. SPSS uses the numbers, not the words, so be careful about keeping all your words and numbers straight.

You must keep accurate records describing your data, how you got the data, and what it means. SPSS can do all the calculations for you, but only you can decipher what it means. In *The Hitchhiker's Guide to the Galaxy,* a computer the size of a planet crunched on a problem for generations and finally came out with the answer, 42. But the people tending the machine had no idea what the answer meant because they didn't remember the question. They hadn't kept track of their input. You must keep careful track of your data or you may later discover, for example, that what you've interpreted to be a simple increase is actually an increase in your rate of decrease. Oops.

SPSS lets you enter the data and tag it to help keep it organized, but you already have the data written down someplace and fully annotated. Don't you?

How SPSS Works

The developers of SPSS have made every effort to make the software easy to use. It prevents you from making mistakes or even forgetting something. That's not to say it's impossible to do something wrong, but the SPSS software works hard to keep you from running into the ditch. To foul things up, you almost have to work at figuring out a way of doing something wrong.

You always begin by defining a set of *variables,* then you enter data for the variables to create a number of *cases.* For example, if you're doing an analysis of automobiles, each car in your study would be a case. The variables that define the cases could be things such as the year of manufacture, horsepower, and cubic inches of displacement. Each car in the study is defined as a single case, and each case is defined as a set of values assigned to the collection of variables. Every case has a value for each variable. (Well, you *can* have a missing value, but that's a special situation described later.)

Each variable is a specific type. That is, each variable is defined as containing a certain kind of number. For example, a *scale* variable is a numeric measurement, such as weight or miles per gallon. A *categorical* variable contains values that define a category; for example, a variable named gender could be a categorical variable defined to contain only values 1 for female and 2 for male. Things that make sense for one type of variable don't necessarily make sense for another. For example, it makes sense to calculate the average miles per gallon, but not the average gender.

After your data is entered into SPSS — your cases are all defined by values stored in the variables — you can easily run an analysis. You've already finished the hard part. Running an analysis on the data is simple compared to entering the data. To run an analysis, you select the one you want to run from the menu, select appropriate variables, and click the OK button. SPSS reads through all your cases, performs the analysis, and presents you with the output as tables or graphs.

You can instruct SPSS to draw graphs and charts directly from your data the same way you instruct it to do an analysis. You select the desired graph from the menu, assign variables to it, and click OK.

When you're preparing SPSS to run an analysis or draw a graph, the OK button is unavailable until you've made all the choices necessary to produce output. Not only does SPSS require that you select a sufficient number of variables to produce output, it also requires that you choose the right kinds of variables. If a categorical variable is required for a certain slot, SPSS will not allow you to choose any other kind. Whether the output makes sense is up to you and your data, but SPSS makes certain that the choices you make can be used to produce some kind of result.

All output from SPSS goes to the same place — a dialog box named SPSS Viewer. It opens to display the results of whatever you've done. After you have produced output, if you perform some action that produces more output, the new output is displayed in the same dialog box. And almost anything you do produces output.

Where SPSS Works

More than one version of IBM SPSS Statistics 18 exists, for execution under different operating systems.

IBM SPSS Statistics 18 for Windows can be run on Windows XP (32-bit) or on Windows Vista (32-bit or 64-bit). You can run IBM SPSS Statistics 18 for Mac on Macintosh 10.5x (Leopard) or on Macintosh 10.6x (Snow Leopard), both 32- and 64-bit. IBM SPSS Statistics 18 for Linux has been tested only on Red Hat Enterprise Linux 5 and Debian 4.0, but it should run on any sufficiently updated Linux system.

All the Strange Words

Statistics seems to have been born in the land of strange words. Lots of them. If you come across a term that you don't understand, such as *dichotomy, variable,* or *kurtosis,* you're not stopped: You can look it up in the glossary at the back of this book.

It's not only new words that can trip you up. You will find common words used in a special way. For example, the word *case* has a special meaning. And a *break variable* has a special purpose when organizing tabular data.

All Those Files

Input data and statistics are stored in files. Different kinds of files. Some files contain numbers and definitions of numbers. Some files contain graphics. Some files contain both.

The examples in this book require the use of files that contain data configured to demonstrate capabilities of PASW. Some of the files are already on your computer, and others can be found on the Internet. Most are in the same directory you used to install PASW. That is, the action of installing PASW also installs a number of data files ready to be loaded into PASW and used for analysis. A few of the files used in the examples can be found in the compressed file `PASW.zip` found on this book's companion Web site (it's listed in the Introduction).

Where to Get Help When You Need It

You're not alone. Some immediate help comes directly from the PASW software package, and other help can be found on the Internet. If you find yourself stumped on some point, you can look in several places, as follows:

 - ✔ **Topics:** Choosing Help⇨Topics from the main window of the PASW application is your gateway to immediate help. The help is somewhat terse, but often it provides exactly what you need. The information is in one large help document, presented one page at a time. Choose Contents to select a heading from an extensive table of contents, choose Index to search for a heading by entering its name, or choose Search to enter a string search inside the body of the help text.

   In the help directory, the titles in all uppercase are descriptions of Syntax language commands.

 - ✔ **Tutorial:** Choose Help⇨Tutorial to open a dialog box with the outline of a tutorial that guides you through many parts of PASW. You can start at the beginning and view each lesson in turn, or you can select your subject and view just that.

✔ **Case Studies:** Choose Help⇨Case Studies to open a dialog box containing examples in a format similar to that of the Tutorial selection. You can select titles from its outline and view descriptions and examples of specific instances of using PASW. You can also find descriptions of the different types of calculations. If some particular analysis type is eluding your comprehension, this is a good place to look.

✔ **Statistics Coach:** Choose Help⇨Statistics Coach if you have a good idea of what you want to do but need some specific information on how to go about doing it.

✔ **Command Syntax Reference:** Choose Help⇨Command Syntax Reference to display more than 2000 pages of references to the Syntax language in your PDF viewer. The regular help topics, mentioned previously, provide a brief overview of each topic, but this document is much more detailed.

✔ **Algorithms:** Choose Help⇨Algorithms to get detailed information on how processes work internally. This is where you can dive far down into the internals. If you want to take a look at the math and how it's applied, this is where you look.

Your Most Valuable Possession

The most valuable possession you have in dealing with statistics is not your computer. It's not your PASW software. It's not even this book, or any other book you may be using to learn statistical procedures. You can lose any one of those, but any one of them can be replaced.

Your most valuable possession is your data. Sure, you can always go and get more data, but you can't go and get the *same* data. The world doesn't hold still long enough. Be sure to make backup copies of your data.

Back up your data to memory that does not live in the same building with the computer you're using. You can swap backups with a friend, or if you have access to a remote Web site, you can stuff files in a blind directory.

This message about backing up your data comes to you from someone who has been stung. And I don't want to talk about it again. Ever.

You Can Dive as Deep as You Want to Go

PASW makes no effort to keep anything secret. It's designed to be as easy to use as possible, so you really don't have to know all that much to make it work. However, if you want to understand how things are working internally, you can find out if you dig. And you don't have to dig very far. Choosing Help is the first step to finding out anything you want to know about what's going on inside.

Let's say you're working on your numbers and want to use some specific algorithm to do your calculations. PASW has been at this longer than you have, so the algorithm you want to use is almost certainly built in. If you're not sure exactly what PASW is doing to calculate some of the numbers, you can go to the Help menu and read through the supplied documentation to find out how the calculations are being performed. But, before you start looking, make sure you really want to know, because the equations and how they are applied are explained in excruciating detail.

The purpose of this book is to give the shallow divers enough information to be able to swim and to show the deeper divers how to begin. I don't explain all the details because there are too many. There's simply not enough room in a book this size to explain PASW in depth.

Chapter 2

Installing Software and Setting Options

This chapter is all about installing your software and setting the options that determine how it works. If the software you'll be using is already installed, you can skip the first part of this chapter and jump right to configuration a little further on. I mention that because it's a *For Dummies* book, and I was told not to leave anything out.

The installation process guides you, step by step, and then does most of the work itself. The configuration settings all default to something reasonable, so I suggest leaving them alone for now. You can always come back later and make a change if you develop a gripe.

Getting SPSS into Your Computer

Soap powder comes in boxes, paint comes in cans, corn dogs come on sticks, and SPSS comes on the Internet. Well, it can also come on a CD, but the installation process is fundamentally the same. The only real difference is where the files come from.

Don't lose track of anything. If you get your software on a CD, find a place to put the package and all its contents. Don't throw out anything. That includes the plastic box in which you found the CD and the cardboard stiffener that came inside the mailing package. If you download the software, keep meticulous records of the Web site, which files you download, all numbers and identifiers you encounter. Trust me, you'll need them later.

The Mac and Linux versions of the software are similar in operation, but details of the installation procedure described here are specific to Windows.

What you need for running SPSS

You won't have to worry about the minimum requirements for the computer — unless yours is an antique. I mean, who *doesn't* have at least 256MB of RAM and 300MB of free disk space?

SPSS comes in a variety of flavors. They're fundamentally alike, but some versions have more parts than others. You may have all, some, or none of the add-ons described in Chapter 20. In any case, you need an authorization code to enable whatever you do have. You will need to authorize your base system as well as any add-ons. You may have more than one authorization code — it depends on how your SPSS system is configured, which is determined by what parts are included with it.

For the installation procedure to work, you must be logged in to your Windows system with administrator privileges. You don't have to be logged in as an administrator, but whatever login you're using must have the privileges that the administrator has.

You should also be connected to the Internet. You can install SPSS without being connected, but it's a pain to do it that way. Make it easy on yourself and connect your computer to the Internet before you start. And keep it connected at least until you get SPSS installed.

In summary, before you begin the installation:

✔ You must have access to your authorization code or codes.

✔ You must have access to the serial number of your copy of SPSS.

✔ You may also need to have access to your customer number.

✔ You must be logged into your computer with administrator privileges.

✔ For convenience, you probably want to be connected to the Internet.

Cranking up the installer

The installation procedure is dead simple: You simply start the installation program and answer the questions. And the questions are easy.

If you have a previous version of the software installed, you have to remove it before you can install the new version. To remove it, use the Windows Control Panel dialog box and select SPSS (or PASW, depending on what you have installed). Then, click the Remove button or Uninstall button to delete it.

You can start the installer in more than one way. The first method is automatic: You insert the SPSS CD into the drive and wait a bit. Most Windows computers will recognize what's on the CD and start the installer. If the installer doesn't start automatically (or if you fool around and close the window after it started), you can execute and then find and run the program called `setup.exe` on the SPSS CD. If you have downloaded your version of the software, you've gotten an executable program. All you need to do is run it. However you do it, you get the window shown in Figure 2-1.

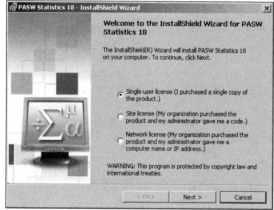

Figure 2-1:
The first window is a list of installation choices.

As you can see in Figure 2-1, you install SPSS according to the type of license you've purchased. The example described in this chapter is for a single-user installation, but you can also install it under a site license or a network license.

The SPSS installation sequence

With the window shown in Figure 2-1 on your screen, select the type of license you have and click the Next button. After you make your selection, you're greeted by the license agreement, as shown in Figure 2-2. Simply do what it says: Read the license, and if you accept the terms, select the I Accept the Terms in the License Agreement option and then click the Next button.

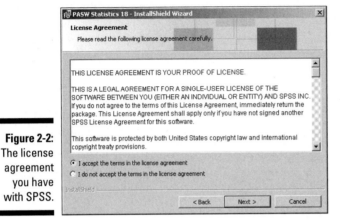

Now the installation gets talky and displays the ReadMe file, as shown in Figure 2-3. If you thought the license was something, wait until you read this stuff. Not all of it will apply to you, but you should read it anyway because you might come across something you need to know, such as where temporary files are stored or the Klingon numbering system. After you've gained all the pleasure you can stand from the ReadMe file, click Next to move on.

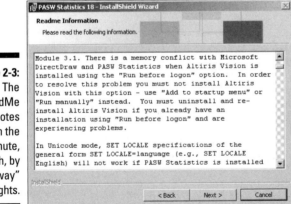

Figure 2-3:
The
ReadMe
notes
contain the
last-minute,
"Oh, by
the way"
thoughts.

The next screen, shown in Figure 2-4, asks for your name and organization. (I always take it as a compliment that the software thinks of me as being organized, but I can never figure out what to put in the blank. You can put anything you like in there, but keep it clean, because it could pop up on the screen one afternoon while your mom is watching.) The third piece of information is a little more important. It wants you to enter the serial number of your copy of the software. This is *not* the authorization code — that comes later. You can find the serial number in two places: on a tag inside the plastic box in which you found the CD and on the cardboard stiffener that came inside the mailing package. If you got your copy online, you should find everything you need somewhere in the e-mails you've received — if not, you may need to request the information.

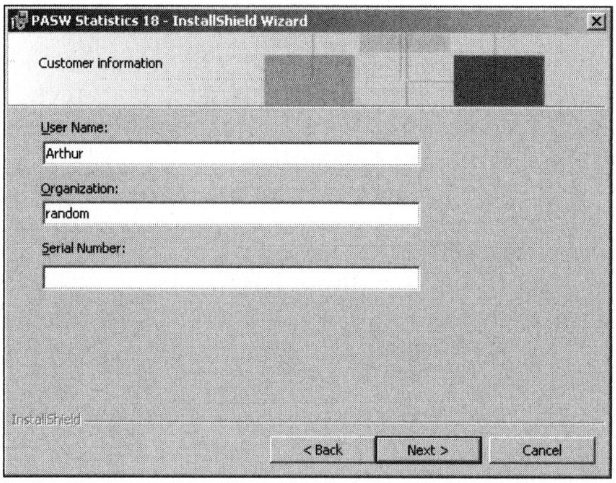

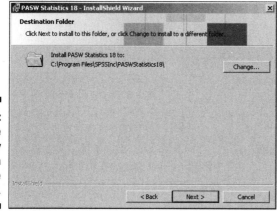

Figure 2-4:
Name, orga-
nization, and
the serial
numbers.

When you click the Next button, you get a window that asks for the directory in which you want SPSS installed, as shown in Figure 2-5. The directory that the installer normally chooses is fine; change it only if you have a really good reason. If you can't think of a reason, accept what's there and move on by clicking the Next button.

Figure 2-5:
The
directory
into which
SPSS will be
installed.

The window that shows up at this point asks you whether you really want to install SPSS. All you've done so far is answer some questions; nothing has been installed. This window has a Back button you can use to go back and change your answers. The Next button unleashes the installation software onto your computer. The screen also has a Cancel button if you chicken out, or if you enjoyed the process so much that you want to drop everything and do the entire thing over again. If you actually want SPSS on your computer, click Next.

The next window, shown in Figure 2-6, lists every file being installed, while a progress indicator moves across the screen. The filenames flicker by pretty fast; only Superman or Data from *Star Trek* could read them. Normal mortals see mostly a line of constantly flickering letters.

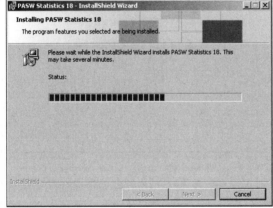

Figure 2-6:
An animated window reassures you that something is happening.

The progress indicator marches across the screen until it reaches the far right. At that point, the flickering of file names will stop. For a time, nothing moves. Be patient. Just about the time you start to wonder whether something has gone wrong, the display presents the window shown in Figure 2-7.

We're coming to the point where it's convenient to be connected to the Internet. You need to select the Register with SPSS.com option shown in the lower-left corner of the window in Figure 2-7. The authorization code is on one of your pieces of paper, or maybe someone else has it. Whether or not you have it, click OK so you will get the product authorization window shown in Figure 2-8.

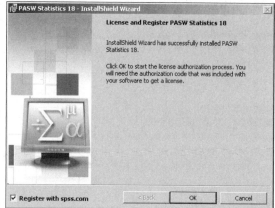

Figure 2-7:
Choose to
move on
to author-
ization.

For full access to the software, you must register it with SPSS.com. If you're
not connected, and you don't register, you can actually get it to work anyway,
but only temporarily. You will have to authorize it (either now or later)
before you can use it permanently. You use this window to select which you
want to do, and then click the OK button.

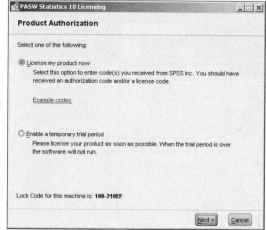

Figure 2-8:
Choose
whether to
authorize
now or
authorize
later.

You can use the window in Figure 2-9, which appears next, to enter your code.
If you think the number you've got is too long, it's probably the right number
(authorization codes *are* long). If you're not sure the number you've got is
an authorization code, you can click the Example Codes link and see what an
authorization code looks like. The code will be sent over the Internet to SPSS.

If you're behind a proxy server, you'll have to provide information that allows a message to get out and a reply to get in, so that SPSS can get a message to come back into your computer.

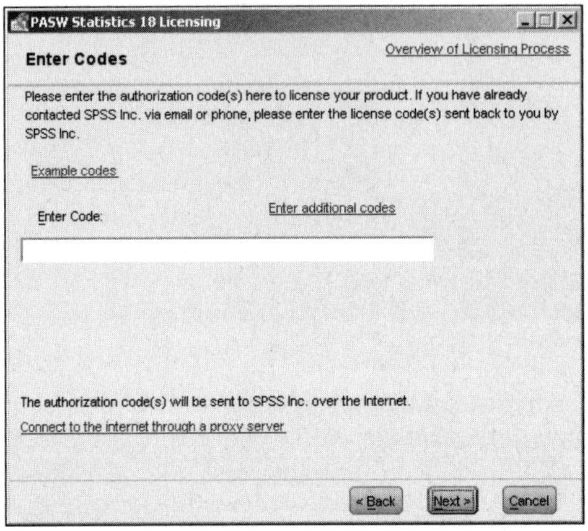

Figure 2-9:
Enter the authoriza-
tion code.

When you're ready, you can enter the code and click the Next button. A window appears with a message verifying authorization. Clicking the Next button brings up the window displayed as Figure 2-10.

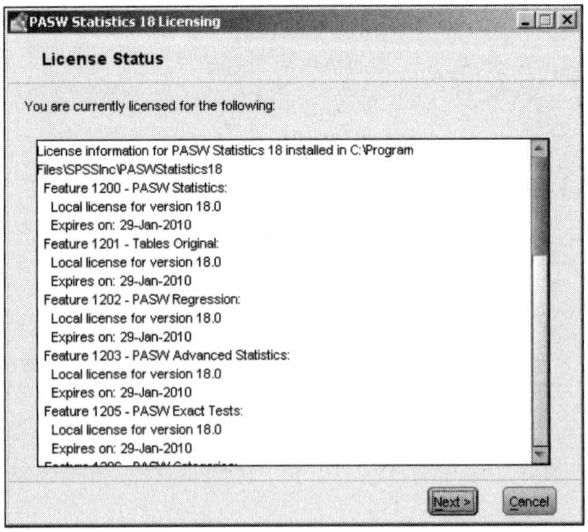

Figure 2-10:
Verification
that the
software
has been
authorized.

An important piece of information included in the window in Figure 2-10 is the expiration date of your license. You'd be well advised to check the dates, and make notes of them for future reference.

Click the Next button. The window disappears and your software is ready to run.

Late registration

If you installed SPSS but chose to register it later, or if you want to check the status of your registration. you can do that easily. Simply select Start⇨Programs⇨SPSSInc⇨PASW Statistics 18⇨PASW Statistics 18 License Authorization Wizard and the status of your license appears on-screen; then you get the same sequence of registration dialog boxes described in the previous section.

The Internet being the Internet, your connection might get dropped right in the middle of the registration process. If that happens, just start over from the Start menu.

Starting SPSS

You now have SPSS installed on your computer. You'll find a listing for it with the other programs on your Start menu. Choose Start⇨Programs⇨SPSSInc⇨ PASW Statistics 18. You then have two choices:

- ✔ PASW Statistics 18
- ✔ PASW Statistics 18 License Authorization Wizard

The first choice is the main program itself — and that will be the number-one selection on your hit parade in days to come. The second choice is the authorization stuff you went through earlier.

When you first start SPSS, you get a window like the one in Figure 2-11. This window makes it possible for you to go directly to the window you want to work with. The problem is that it assumes you already know what you want to do, but so far you have no idea what you want to do with SPSS yet, so just click the Cancel button to close the window.

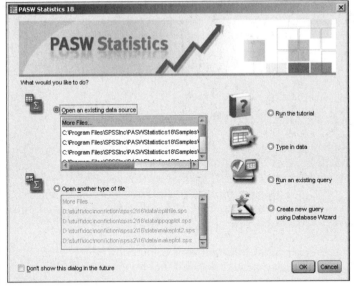

Figure 2-11:
You can go
directly to
the function
you want to
perform.

You see the regular Data Editor window, shown in Figure 2-12. If you've ever worked with a spreadsheet, this display should look familiar. And it works much the same way. This window is the one you use to enter data. I generally like to expand the window to fill the entire screen because more spaces are displayed at one time. Besides, I don't need to see any other windows because I almost never do two things at once.

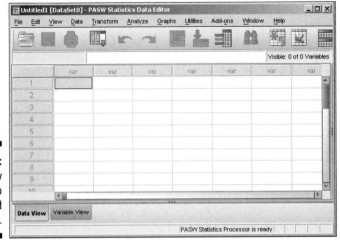

Figure 2-12:
The window
used to
enter and
view data.

Exploring and Modifying the Default Settings

Over time, you'll find that you want to configure your system to work in ways that are different from the defaults. SPSS has lots of options that you can set to do just that. If you're new to this and have just started looking at the software, you probably don't want to change any options just yet, but you need some idea of the possibilities it offers. Later, when you absolutely, positively have to make some sort of change, you'll know where to go to do it.

With the Data Editor on the screen (refer to Figure 2-12), choose Edit⇨Options to display the Options window. You can set any and all possible options in the Options window. At the top of this window are some tabs; each tab selects a different collection of options. Sometimes a change in configuration has an immediate effect, and sometimes it doesn't. For example, if you change the way values are labeled in a report that's already on-screen, nothing happens because the report has already been constructed. You have to run the report-generating software again — so it builds a new report — to have the changes take effect.

General options

The first tab in the Options window, the General tab (shown in Figure 2-13), displays a dialog box with options that don't fit into any of the categories defined by the other tabs.

The options displayed by the General tab are as follows:

- ✔ **Variable Lists:** Lists of variables in your output can be identified by either their labels or their names. You can think of these as short titles and long titles, and you can have your data items, by default, tagged by one or the other as they appear in lists. Also, you can have your data appear in alphabetical order by the titles you use for individual items, or simply by the order in which the data appears in the file. File order usually makes more sense.

- ✔ **Roles:** When you select some actions, variables of the types that can play certain roles in the processing to follow can be preselected for you if you have the first option (Use predefined roles) selected. If you have the other option (Use custom assignments) selected, you will be required to choose all the variables yourself.

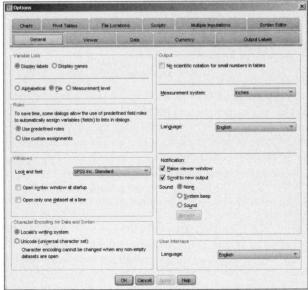

Figure 2-13:
The General
options are
about the
form of out-
put data.

✔ **Windows:** These are cosmetic options. You can choose how you like the dialog boxes of SPSS to appear.

• *Look and Feel*: Your choices are SPSSInc Standard, SPSSInc Classic, and Windows.

• *Open the Syntax Window at Start-up:* Makes SPSS begin with the syntax window instead of the data editor. Choose this option if you use the scripting language more often than the windowing interface to enter data and run your predefined procedures.

• *Open Only One Dataset at a Time:* Whenever you open a new dataset, the new information appears in a new window and any that are already open are closed. With this option set, the already open dataset is closed when the new one is opened. By the way, this does not apply to datasets opened inside a syntax language process.

✔ **Character Encoding for Data and Syntax:** You can choose to read and write files in Unicode mode, but you shouldn't unless you have a good reason to do so. The files are read and written in UTF-8 format. If you write a Unicode file, you need to be sure that the software that reads it understands that format. When you read a file in Unicode mode, it's much larger in memory than it would be otherwise.

✔ **Output:** These options control the appearance of tables and graphs:

- *No scientific notation for small numbers:* Suppresses scientific notation for small numbers. For example, 12 appears as 12 instead of 1.2e1, which is a little harder to read. SPSS doesn't say exactly what it considers to be a small number.

- *Measurement System:* Units used to specify the margins between table cells, the width of cells, and the spacing between printed characters. You can use inches, centimeters, or the default, points. (A point is ½ of an inch.)

- *Language:* The output language can be set to any one of about a dozen choices, and it determines the language used to output files. You may have use Unicode mode to handle all the characters in some languages.

- *Notification:* The method the software uses to notify you when the results of a calculation are available. With the Raise Viewer Window option, the display window opens automatically. With the Scroll to New Output option, the window scrolls and exposes the location of the new data. You also can have the system beep, tweet, or sing when an analysis is complete. (It's considered impolite to have it make rude comments when an analysis finishes.)

✔ **User Interface:** The user-interface Language setting determines the language used to display menus and dialog boxes. Life is so much easier if you choose a language that you actually know how to read.

Viewer options

Output from SPSS is formatted for viewing with either the draft viewer or the regular viewer. SPSS thinks in terms of a printed page, but the same layouts are used for displaying data on the screen. The options you can set for the regular viewer can be accessed with the Viewer tab, shown in Figure 2-14.

The options in the Viewer tab are as follows:

✔ **Initial Output State:** Determines which items are displayed each time you run a procedure. You choose an item by either selecting its name (Log, Warnings, Notes, Title, and so on) from the pull-down list or by selecting its icon. Then you can select whether you want it to appear or remain hidden, how you want its text justified (Align Left, Centered, or Align Right), and whether the information occurrence should be included as part of the log (Display Commands in the Log).

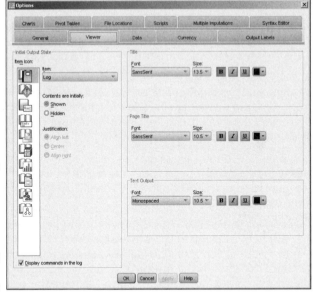

✔ **Title:** Here you choose the font used for main output titles. It appears at the top of the first page of a report.

✔ **Page Title:** Choose the font used for the title appearing at the top of subsequent pages of a report.

✔ **Text Output:** Determines the font used for the text of your report and the labels on graphs and tables. The font size also affects the page width and length because the sizes are measured by counting characters.

Some fonts have variable-width characters, which will throw off the alignment of your columns. If you want everything to align in neat columns all the time, use a monospace font.

Data options

The Data tab, shown in Figure 2-15, can be used to specify how SPSS handles a few special numeric situations. The options in the Data tab are as follows:

✔ **Transformation and Merge Options:** Determines when — not how — results are calculated. You can have SPSS perform calculations immediately, or you can have it wait until it needs the number for something (either another calculation or a displayed value). Both methods have their advantages and disadvantages.

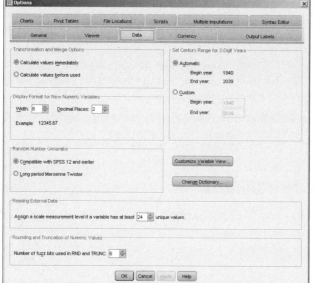

Figure 2-15:
Data
handling in
SPSS can
be varied
within the
limits of the
settings of
this tab.

✓ **Display Format for New Numeric Variables:** Determines how many digits are used in the display of values, and how many digits are to the right of the decimal. Width is the total number of characters, including the decimal point. The Decimal Places setting determines the number of digits that appear to the right of the decimal point. If the number of places to the right is too small, values are rounded to fit. If the number of places is too large, values are put in scientific notation.

✓ **Random Number Generator:** Ever since a need for random numbers was discovered, generating them on a computer has been a problem — because computers naturally do things in a *non*-random way. SPSS offers you two ways to do it: the old way and the new way. If you'd rather not generate your random numbers the same way you did in older versions of SPSS (version 12 and earlier), use the Twister.

✓ **Reading External Data:** When SPSS reads numeric data, it counts the number of unique values assigned to a variable and uses the count to determine whether the variable is nominal or scalar. The count you enter here determines the threshold used to make that determination.

✓ **Rounding and Truncation of Numeric Values:** This setting determines the threshold for rounding numbers. SPSS does the calculation in base two, so the count is a number of bits. *Fuzz bits* refers to a count of the number of bits to be considered.

✓ **Set Century Range for 2-Digit Years:** A solution to the Y2K problem. I'll bet you thought that was all over. It is, but the solutions are still with us — and this is one of them: You put in two four-digit years here, and any two-digit value that you supply to identify a year is assumed to be between the two years you specify. This is mostly for old data. If you always use four digits for years in your data, this century range setting is never used.

✔ **Customize Variable View button:** Allows you to determine which variable attributes are displayed, and in what order they are displayed, in the Variable View window of SPSS.

✔ **Change Dictionary button:** Allows you to determine which dictionary is used to check the spelling in Variable View.

Currency options

Different parts of the world use different symbols and formats when writing about currency. The window shown in Figure 2-16 lets you specify the display format of your currency.

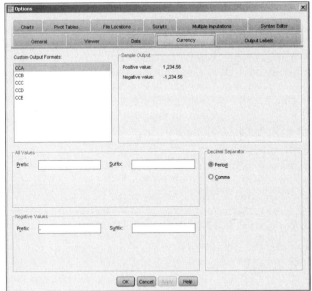

Figure 2-16:
Number formats and symbols can be set so SPSS displays things correctly for your currency.

Following are the options in the Currency tab:

✔ **Custom Output Formats:** The default format for presenting currency values. The five formats have the unlikely names CCA, CCB, CCC, CCD, and CCE. Those are the only ones you can have, but that has to be enough for anybody (I mean, really, if you work with more kinds of money than that, buy another copy of SPSS). The calculations are always performed the same way — the differences are in the display. You can set the display configuration for each one to be anything you'd like (dollars, euros, yen, and so on), and then switch among them as often as you wish.

✓ **Sample Output:** Displays the printed format of positive and negative currency values. As you switch from one currency selection to another, and as you change the formatting of any of them, the sample displays examples of the format.

✓ **All Values:** Specifies the characters that appear on-screen to identify the currency, at the front or back of all values. Such characters include the British pound sign and the cent mark.

✓ **Negative Values:** Specifies the characters placed in front and in back of negative values. For example, some folks like to use < and > to surround negative money values.

✓ **Decimal Separator:** Many currency notations use commas instead of periods to denote the fractional portion of the amount.

Output Labels options

Every variable can be identified in two ways: by a label and by a name. In your output, you can specify to have variables identified by one or the other or both. You configure output labeling on the Output Labels tab, shown in Figure 2-17.

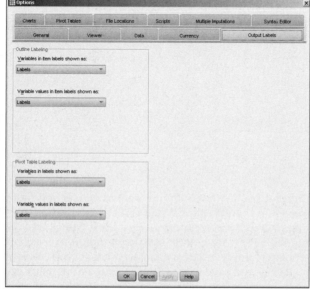

Figure 2-17: Here you specify whether to include names, labels, or both when you label your data.

With these options, you can choose to display the variable names, the variable labels, or both. Longer labels can be descriptive and make your data easier to determine, but they can also screw up some formats. Following are the options in the Output Labels tab:

- ✔ **Outline Labeling:** The text used to identify the parts of charts and graphs.
- ✔ **Pivot Table Labeling:** The text used to identify the rows and columns of tables.

Chart options

The default appearance of charts is determined by the settings in the Charts tab, shown in Figure 2-18.

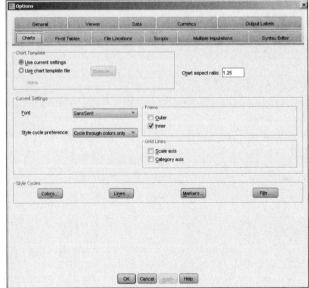

Figure 2-18:
Change
the default
appearance
of a chart
with these
settings.

The options in the Chart tab are as follows:

- ✔ **Chart Template:** A file that contains a set of starter settings that you can use for designing a new chart. When you create a new chart, it can use the settings in this configuration window, or it can use this file. You can select any file to be your default starting template. It's easy to create a chart template: Simply create a chart that has all the configuration settings you like — and save it so you can use it as the template file.

✔ **Current Settings:** This section offers two pull-down menus, as follows:

- *Font:* The default font for the text in any chart you design.

- *Style Cycle Preference:* How SPSS chooses the styles and colors when laying out data items in a chart. You can have SPSS cycle through just the colors so each item included in the graph is identified only by its color. If you're using a black-and-white printer or display, choose Cycle Through Patterns Only: Each data item is identified by a graphic pattern of line styles and marker symbols.

✔ **Style Cycles:** Customizes the sequence of colors and patterns to be cycled through.

✔ **Chart Aspect Ratio:** The ratio of the width to the height of the produced charts, initially set to 1.25. Which ratio looks better is a matter of opinion; you'll have to experiment.

✔ **Frame:** Determines whether charts display an inner frame, an outer frame, both, or neither.

✔ **Grid Lines:** Displays dividing lines on the scale axis, on the category axis, or on both.

Pivot Tables options

The tabular output format of SPSS is the *pivot table.* An example is shown as the sample in Figure 2-19, which is the Pivot Tables tab used to set display options for the tables.

The options in the Pivot Tables tab are as follows:

✔ **TableLook:** A file that contains your standard pivot table and determines the initial appearance of any new tables you create. Several such files come with the system and are listed in the window. You can also create your own file by choosing TableLook from the menu in the Pivot Table Editor window. The Set TableLook Directory button sets the currently displayed directory as the one in which your new table files are stored. You can choose any directory you like; clicking this button causes your chosen directory to appear in this window by default.

✔ **Column Widths:** Controls the way SPSS adjusts column widths in pivot tables. You can adjust them according to the width of the labels or according to the width of the data or labels, whichever is wider.

✔ **Display Blocks of Rows:** These settings determine the size at which pivot tables appear on-screen in the Viewer window. You can set the number of rows to display in each section and the maximum number of cells to display in each section. The widow/orphan setting has to do with the number of categories that must appear before and after a row is split.

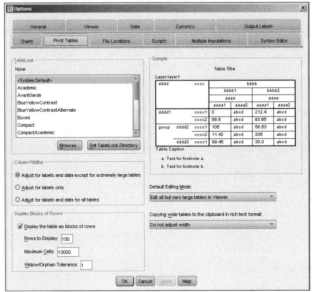

Figure 2-19:
The options
you set in
this window
determine
the appear-
ance of a
new table.

✔ **Default Editing Mode:** When you double-click a pivot table, you can edit it in place or a separate edit window is opened, depending on this setting.

✔ **Copying Wide Tables to the Clipboard in Rich Text Format:** When a table is copied to the Word format or Rich Text Format, tables too wide for the document are wrapped to fit, scaled to fit, or left as they are, depending on what you choose for this setting.

File Locations options

The options on this tab specify the locations of the files opened for input and output. The option settings are shown in Figure 2-20.

The options for the file locations are as follows:

✔ **Startup Folders for Open and Save Dialogs:** The startup folders are the names of the directories that initially appear in the Save and Open dialog boxes when you read or write data files. Optionally, you can select to simply use the last directory used to read or write a file.

✔ **Session Journal:** You have the option to configure a journal file to receive a copy of every Syntax language command, whether it comes from a script or from a user entering instructions through a dialog box.

✔ **Temporary Folder:** You can specify the name of the directory where SPSS creates its temporary working files.

✔ **Number of Recently Used Files to List:** The most recently read or written files are listed in the Files menu. This option specifies how many are listed.

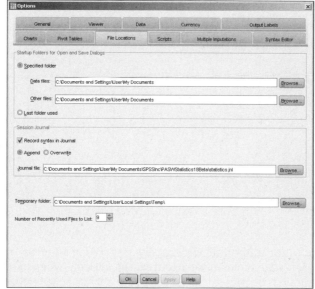

Figure 2-20:
The options
you set in
this window
determine
the appear-
ance of a
new table.

Scripts options

Figure 2-21 displays the Scripts tab, which is used to determine some fundamental defaults about scripts.

Don't mess with any of these until you've been writing scripts for a bit and know what you're doing because a single change here can affect the execution of a number of scripts:

✔ **Default Script Language:** This setting determines which script editor is launched when new scripts are created. The default script language is Basic. No other choice is available unless you have the Python add-on installed.

✔ **Enable Autoscripting:** If you choose this option, you enable the autoscripting feature.

✔ **Base Autoscript:** A script, if it's stored in the file you name here, defines a global procedure that runs automatically when you create an object. It always runs before any other autoscript for that object. The choice of languages for it can be either Basic or Python (and then only if the Python add-on is installed).

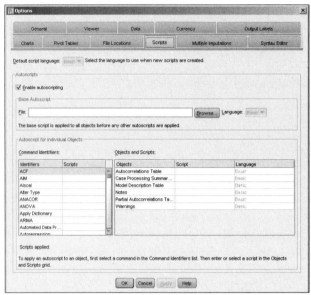

Figure 2-21:
This window
configures
global pro-
cedures and
scripts that
run auto-
matically.

✔ **Autoscript for Individual Objects:** By associating a type of object with an autoscript, you can make an autoscript execute when an object of that specific type is created. To associate an autoscript with an object type, first select the command that generates an object of the desired type (these commands appear in the Identifiers column on the left). On the right, the Objects column then displays the types of objects that your chosen identifier command will generate. In the script cell to the right of the object type you want to tag, enter the path name of the file containing the autoscript. Alternatively, you can click the ellipsis button that appears in the cell and browse for a script file. When you've chosen the file you want, click the OK or Apply button to make the association.

To delete an autoscript association, in the Script column on the right, select the name of the script file you wish to disassociate, and then delete it. Select some other cell to make sure your deletion has been accepted, and then click the OK or Apply button.

Multiple Imputations options

SPSS keeps track of which data has been entered and which has been *imputed* (assumed). The imputation process is that of calculating what the values of your missing data *would be*. You can set multiple imputation options using the window shown in Figure 2-22.

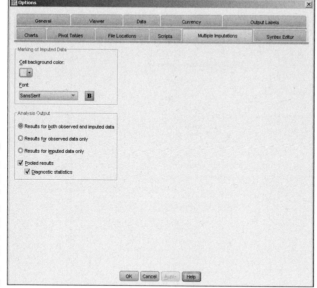

Figure 2-22:
Here you can specify how the imputed data is displayed and handled.

Following are the options in the Multiple Imputations tab:

- ✔ **Marking of Imputed Data:** You can change the appearance of imputed data in the Data View display. You can highlight it by changing the background color of the cell in which it's displayed, and by using a different font to write its values.

- ✔ **Analysis Output:** You can choose to do analytical calculations using imputed data, without using it, or both ways. Also, you can set the imputation process to pool previously imputed data for further imputation. I'd suggest leaving this setting alone for now — it takes a hairy-legged mathematician to figure it all out. Some analysis procedures can produce separate analysis results using only imputed data — you can choose to generate output from such pooled data.

Syntax Editor options

The editor of the Syntax Command language is capable of recognizing various language parts and highlighting them for you. The Syntax Editor options window is shown in Figure 2-23.

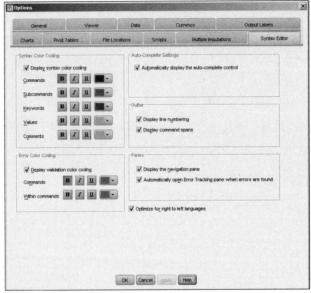

Figure 2-23:
You can
specify how
you'd like
the editor
to display
Syntax
code.

Following are the options in the Syntax Editor tab:

- ✔ **Syntax Color Coding:** You can specify different colors for commands, subcommands, keywords, values, and even comments. You can also specify each one as bold, italic, or underlined. A single switch turns on all coloring and highlighting.

- ✔ **Error Color Coding:** You can specify the font style and the color coding of error information. A single switch turns on all coloring and highlighting.

- ✔ **Auto-Complete Settings:** Use this switch to suppress or allow the display, in the Syntax Editor window, of the option button that turns auto-complete on or off.

- ✔ **Gutter:** The space to the left of the commands is called the gutter. Various types of information are displayed there. You can use the gutter to display the line numbers or the span of a command (the beginning and ending of a single command).

- ✔ **Panes:** You can display or hide the navigation pane, which contains a list of all Syntax commands. You can also cause the error-tracking pane to automatically appear when SPSS encounters an error.

- ✔ **Optimize for Right to Left Languages:** You have to select this option when you're working in a language that reads right to left.

Chapter 3

A Simple Statistical Analysis Example

The purpose of this chapter is to introduce you to the mechanics of working with SPSS. It begins with stepping through the process of entering some simple data into SPSS and continues with processing that data. This is followed by various procedures for deriving results, using a subset of the data for some calculations and then some other parts of the data for other calculations. Finally, the results from these different calculations are displayed in different ways.

The data for this example is simple, as are the displays that the data generates. The purpose of this chapter is not to present any great breakthrough in statistical analysis. Instead, I simply want to demonstrate the basic procedures you need to know about when you're using SPSS.

When the Tanana at Nenana Thaws

This analysis is about an annual lottery that takes place in Alaska. Actually, it isn't called a lottery — it's called a *classic*, whatever that means.

I don't know whether the Tanana Classic is the oldest lottery in the United States (it began in 1917), but it's certainly the slowest. It has only one jackpot per year, and tickets for that jackpot are sold all across the state during the winter months.

The lottery is simple enough: The citizens of the town of Nenana set up a large tripod on the ice in the middle of the Tanana River. From the top of the tripod, a tight line is stretched to a clock on a bridge. When the spring thaw comes, the tripod moves and the clock is triggered, stamping the exact minute. All the people who have selected the correct month, day, hour, and minute share the pot.

Many questions come to mind. What is the most likely date? What is the most likely time of day? Is there a trend? In the analysis that follows, we'll look at the answers to these questions and more.

By the way, the earliest the ice moved out was April 20 at 3:27 p.m. (in 1940), and the latest was May 20 at 11:41 a.m. (in 1964).

Entering the Data

SPSS can acquire data from many sources. You can instruct it to read data from a text file, a database, or a file produced by a program such as Access or Excel. This Alaskan example does it the simplest way possible: by typing data into the Editor window of SPSS. (I said *simplest*, not easiest.)

The data consists of dates and times. SPSS has a special date format that we'll be using later, but for now, we'll enter the year, month, day, hour, and minute as separate numeric items. This keeps the example as simple as possible, and enables me to show you some different ways of manipulating numbers to reach conclusions.

Entering the data definitions

The first job is to define the names, labels, and data types for the various fields of data, also known as the *variables*. Here's all you need to do:

1. **Start the SPSS program by choosing Start⇨Programs⇨SPSS Inc⇨PASW Statistics 18⇨PASW Statistics 18.**

    Depending on how your software is configured, you may get an options window with OK and Cancel buttons. If so, click the Cancel button. In either case, an empty Data Editor window appears, as shown in Figure 3-1.

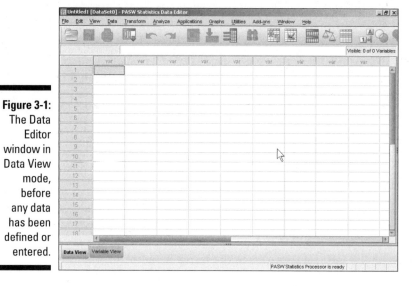

The layout shown in Figure 3-1 is the Data View mode, as indicated by the tab at the bottom of the window. We want to go to the other mode.

2. Click the Variable View tab.

The window now looks like the one in Figure 3-2.

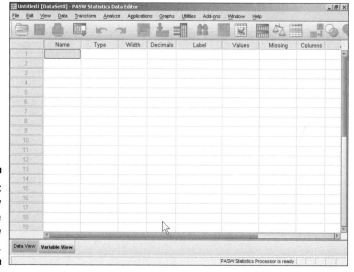

You use the Variable View window to define the names and types of variables, and you use the Data View window to enter the values for those variables.

To enter the definitions, you type the name in the first column — the one labeled Name at the top — and then move the cursor down one row to the position for the next name in the list. You can most easily move the cursor by clicking the destination cell with the mouse. You can also move the cursor with the Enter key and the arrow keys, but the movement may not always be in the direction you expect.

In Figure 3-3, I entered the variable definitions we use in this example.

When you move down to define a new variable name, SPSS takes a wild guess at what you want in the cells you skipped and fills them in for you automatically. Some of the guesses are right, and some are wrong. Stick with me here, and I'll describe the fiddling around you have to do until your information matches that in Figure 3-3.

 3. **Type the following entries in the Name column:**

   • year

   • month

   • day

   • hour

   • minute

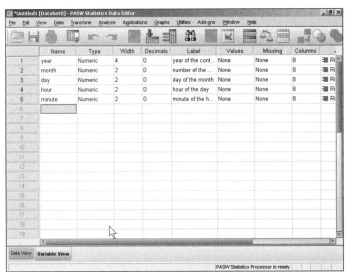

Figure 3-3:
Definition of
the variable
names.

Every field has both a name and a label. One or the other is used as an identifying tag when data is displayed.

The name is normally shorter than the label. A short name is handy when you're displaying data in a tight format, such as a column heading or a bar chart label, and when you're writing equations in the two scripting languages supplied with SPSS. The label is intended to be more descriptive and can add clarity by being displayed as descriptions in displays such as line graphs and pie charts.

4. Skip the Type column.

In this example, all the fields are simple numerics, so SPSS guesses correctly about most of the attributes and fills them in for you. Most of the data you enter into SPSS will be numeric, although some numbers will be converted into names by SPSS. It's hard to perform calculations with things like "moonbeam" and "sure bubba," but it can be done. Later on, I'll be showing you how to instruct SPSS to change numbers into words and phrases automatically.

SPSS has set the number of digits to the right of the decimal point (the Decimals column) to 2 for all the numbers in our example, but that's not what we want for this example.

5. Set all the values in the Decimals column to zero.

This has to be done before you can adjust the Widths — if you don't believe me, try it.

6. Set the first value in the Width column to 4 and the rest of the values in that column to 2.

The Width of most of the fields in this example should be changed from the default of 8 to 2 because they're two digits long. But set the year to a Width of 4 to accommodate 4 digits (we don't want to do the Y2K thing all over again). Simply click the box (cell) for the year's Width column and type 4.

By the way, SPSS has a nifty `date` data type. I didn't use it here because I want to show you how to work with simple numbers. You find out about dates and some other special types and formats in Chapter 4.

7. Type the following into the cells in the Label column:

- Year of the contest

- Number of the month

- Day of the month

- Hour of the day

- Minute of the hour

When you type the Label, you're not limited to the size of the cell that holds it. If you type a longer line, the box expands to take it all in. But don't write a thesis; you need something that will display nicely on your graphs and tables. (You can always come back and change it later on.)

Depending on how big you've made your window, you may have to scroll to display columns to either side. To scroll, use the horizontal scroll bar at the bottom of the screen. (I like to expand my window to the full screen, but that's probably because I'm easily distracted if I see other windows.)

8. **Skip the Values column; it's for assigning names to specific values, and isn't used until later in this example.**

9. **In the column labeled Missing, specify whether it's okay to have values missing from this field.**

For example, if you are taking a survey on what color underwear people are wearing, you could assign a number to each color, but you are bound to come across someone who isn't wearing any, so you'll need to define a special value used to indicate a missing item. By default, SPSS does not allow for missing data, and this example doesn't have any, so the default is None.

10. **Skip the Columns column.**

The default column width for a data item is 8, and that's okay for this example. You can make the columns smaller, if you prefer, but you need to make sure the columns are big enough to hold your largest data item or its name. This is the amount of space that SPSS allocates when it constructs charts and tables. If you set the size too small, the data or the variable name will be cut for some displays.

11. **In the Align column, specify the alignment of your data.**

You can choose whether the data should be aligned on the right, shoved over to the left, or placed in the center. Choose whatever you like. This is determined by personal preference, a lousy sense of design, and bad taste.

12 **The next to the last column on the right is labeled Measure. It can be set to Scale (the default), Ordinal, or Nominal. Leave it set to Scale.**

Scale is an amount or size — it's just a regular number — and works fine for what we're going to do. Ordinal has to do with things that have a specific order. Nominal values are used to tag things as belonging to categories.

13. **Skip the Role column.**

The Role of all variables in this example is standard: they hold input data. They could also be tagged as Target (or output) data, or as Both, or as None. A variable can be also designated as Partition and used to divide the data into separate samples.

Entering the actual data

Click the Data View tab, which is at the bottom of the Data Editor window, and the window changes to look like the one shown in Figure 3-4. The label names you entered in the Variable View window appear at the top of the columns. This window is now ready for you to enter numeric data.

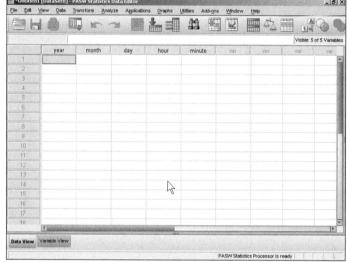

Figure 3-4:
The Data View window ready to accept your input.

In Figure 3-4, notice the numbers down the left side of the window. This is the SPSS way of numbering rows, which are also called *cases*. If you use the scroll bar on the right side of the window to scroll down, you'll see these numbers change. You can think of these numbers as a roadmap to the layout in the window so you can keep track of where you are.

However, don't trust the numbers to identify your data. If you move your data from place to place in the grid, the numbers on the left don't move with it. That means if you insert a row, delete a row, or simply sort your data in a different way, the numbers on the left will associate with different sets of values — and your case numbers will all be different. If you need to identify a case in a manner that does not change when someone organizes the cases differently, then you must add a field for identity and enter your own identifying numbers.

All the values that must be entered for this example are in the following list — but you can be lazy if you want to; I've already entered all the numbers. All you have to do is load the file that holds them by choosing File➪Open➪Data and then selecting nenana.sav. But even if you decide to read them in from the file, enter a few anyway so you can see how SPSS data entry works. (I talk about loading the file a little later.)

4/20/1940, 15:27	4/30/1951, 17:54	5/5/1963, 18:25	5/10/1931, 09:23
4/20/1998, 16:54	4/30/1978, 15:18	5/5/1987, 15:11	5/10/1972, 11:56
4/23/1993, 13:01	4/30/1979, 18:16	5/5/1996, 12:32	5/10/1975, 13:49
4/24/1990, 17:19	4/30/1981, 18:44	5/6/1928, 16:25	5/10/1982, 17:36
4/24/2004, 14:16	4/30/1997, 10:28	5/6/1938, 20:14	5/11/1918, 09:33
4/26/1926, 16:03	5/1/1932, 10:15	5/6/1950, 16:14	5/11/1920, 10:46
4/26/1995, 13:22	5/1/1956, 23:24	5/6/1954, 18:01	5/11/1921, 06:42
4/27/1988, 09:15	5/1/1989, 20:14	5/6/1974, 15:44	5/11/1924, 15:10
4/27/2007, 15:47	5/1/1991, 12:04	5/6/1977, 12:46	5/11/1985, 14:36
4/28/1943, 19:22	5/1/2000, 10:47	5/6/2008, 22:53	5/12/1922, 13:20
4/28/1969, 12:28	5/2/1960, 19:12	5/7/1925, 18:32	5/12/1937, 20:04
4/28/2005, 12:01	5/2/1976, 10:51	5/7/1965, 19:01	5/12/1952, 17:04
4/29/1939, 13:26	5/2/2006, 17:29	5/7/2002, 20:27	5/12/1962, 21:23
4/29/1953, 15:54	5/3/1919, 14:33	5/8/1930, 19:03	5/13/1927, 05:42
4/29/1958, 14:56	5/3/1941, 13:50	5/8/1933, 19:30	5/13/1948, 11:13
4/29/1980, 13:16	5/3/1947, 17:53	5/8/1959, 11:26	5/14/1949, 23:39
4/29/1983, 18:37	5/4/1944, 14:08	5/8/1966, 12:11	5/14/1992, 06:26
4/29/1994, 23:01	5/4/1967, 11:55	5/8/1968, 21:26	5/15/1935, 13:32
4/29/1999, 21:47	5/4/1970, 10:37	5/8/1971, 21:31	5/16/1945, 09:41
4/29/2003, 18:22	5/4/1973, 11:59	5/8/1986, 22:50	5/20/1964, 11:41
4/30/1917, 11:30	5/5/1929, 15:41	5/8/2001, 13:00	
4/30/1934, 14:07	5/5/1946, 16:40	5/9/1923, 14:00	
4/30/1936, 12:58	5/5/1957, 09:30	5/9/1955, 14:13	
4/30/1942, 13:28	5/5/1961, 11:30	5/9/1984, 15:33	

You should now be displaying the Data View window. To enter a number, simply click a position with the mouse and then type the number that you want to put in that square.

When I entered the data, I duplicated a row that was already there and then made changes to it. This was handy because the month and day of the new entry were often the same as the duplicated entry. To duplicate a row, select the row you want to copy by clicking the number at the left of the row. (One click selects the entire row.) Then choose Edit⇨Copy. Next, select the row you want to hold the duplicate data, and then choose Edit⇨Paste. If your target row already contains data, the new data overwrites it.

Suppose you want to insert a new row of data in front of some you already have. First, select the row that is in the position you want to insert the new row; then choose Edit⇨Insert Cases to open a blank row in the position you have chosen. You can either copy or type new data into the blank row.

When you're finished, you can scroll up and down and see different parts of the data, as shown in Figure 3-5.

	year	month	day	hour	minute	var	var	var	var
70	1923	5	9	14	0				
71	1955	5	9	14	13				
72	1984	5	9	15	33				
73	1931	5	10	9	23				
74	1972	5	10	11	56				
75	1975	5	10	13	49				
76	1982	5	10	17	36				
77	1921	5	11	6	42				
78	1918	5	11	9	33				
79	1920	5	11	10	46				
80	1985	5	11	14	36				
81	1924	5	11	15	10				
82	1922	5	12	13	20				
83	1952	5	12	17	4				
84	1937	5	12	20	4				
85	1962	5	12	21	23				
86	1927	5	13	5	42				

Figure 3-5: The data freshly entered into SPSS.

When you're entering your own data, select a filename early in the process and choose File⇨Save to write everything to that file from time to time. If you don't do this, a simple computer crash could lose all your data. That sort of thing is not good for your blood pressure.

By the way, if you've scrolled all the way down, you've noticed that there's a bottom to the list of numbered rows. Don't worry about it. As you enter data, the bottom extends so you never hit a limit.

If you've elected not to enter the data by hand, and instead want to load it from the file, choose File⇨Open⇨Data, and then navigate to wherever you stored the nenana.sav file, as shown in Figure 3-6. Depending on how your Windows system is configured, the name may be chopped off in your display and appear only as nenana. It's not abnormal for Windows to change filenames this way. The book's Introduction tells you how and where you can get the files.

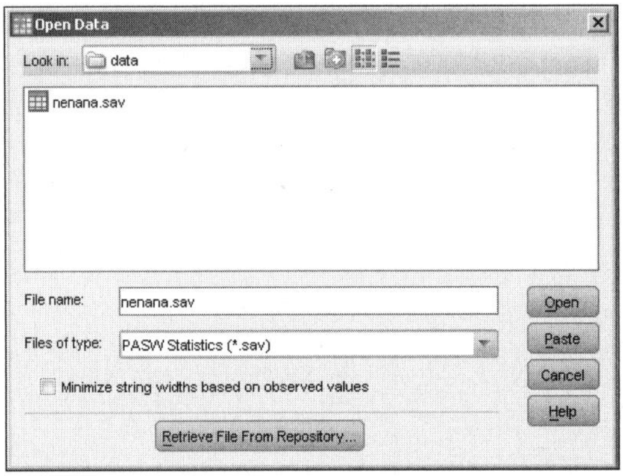

Figure 3-6: Loading an SPSS data file.

The Most Likely Hour

After you've put the data in SPSS, do something simple. Use the following procedure to find the mean of the hours in an attempt to determine the hour of the day when the ice is most likely to melt. This would probably be in the daytime because the sun is warming both the air above the ice and the flowing water below the ice.

To find the most likely hour (ignoring the minutes for now), follow these steps:

1. **Choose Analyze⇨Descriptive Statistics⇨Descriptives.**

2. **In the box on the left, select** hour of the day **(one of your variable labels) and then click the arrow button in the middle of the window.**

   The label moves to the right, as shown in Figure 3-7.

3. **Click the Options button.**

4. **Select the Mean, Std deviation, Minimum, and Maximum check boxes, as shown in Figure 3-8.**

5. Click Continue.

6. Click the OK button at the bottom-left of the window in Figure 3-6.

The SPSS Statistics Viewer appears and displays information about the analysis, including the results. A detailed description of all this information is in Chapter 8. For now, expand the window to fill the screen, and use the scroll bars if necessary, to locate the result in the box at the bottom of the right panel of the window, as shown in Figure 3-9. The mean (not the average, but nearly the same thing) shows the hour as 14.68, which is in the afternoon between 1 and 2 o'clock. That makes sense, because that's near the warmest part of the spring day.

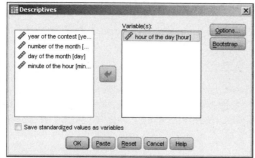

Figure 3-7:
Selecting
data and
starting the
analysis.

Figure 3-8:
The option
settings for
the analysis.

Figure 3-9:
The results
of the
simple hour
analysis.

Descriptive Statistics

	N	Minimum	Maximum	Mean	Std. Deviation
hour of the day	92	5	23	14.68	4.098
Valid N (listwise)	92				

Inside the box, the text on the far left is the label you gave to the variable. The column labeled N is the number of data items included in the calculations. You can tell from the minimum and maximum that the earliest the ice has ever let go was during the 5 o'clock hour in the morning, but it has also been known to happen after 11 at night.

The value for the standard deviation is calculated according to the degree of variation from a perfect fit on a bell curve.

There's more bell-curve stuff to diddle with: Go back through the same procedure again, but this time change the options in Step 4. Remove maximum and minimum and instead enable Kurtosis and Skewness. Those are not rude words (and, no, I didn't just make them up); they're types of statistics. As shown in Figure 3-10, the results have two new values.

Figure 3-10:
New
analysis
showing
kurtosis and
skewness.

Descriptive Statistics

	N	Mean	Std. Deviation	Skewness		Kurtosis	
	Statistic	Statistic	Statistic	Statistic	Std. Error	Statistic	Std. Error
hour of the day	92	14.68	4.098	.085	.251	-.495	.498
Valid N (listwise)	92						

Both values also have to do with the bell curve. *Skewness* represents the symmetry of the data. A positive skewness indicates that more of the data appears to the high end, or the right, on the graph. A negative value indicates a skew to the lower values. *Kurtosis* has to do with the flatness of the curve. If the data implies a curve flatter than the bell curve, the kurtosis value is negative. If, on the other hand, the data inscribes a curve that is more pointed on top than the bell curve, the kurtosis value is positive.

Transforming Data

The previous example looks at only the hours, but it's also possible to include minutes. Clock arithmetic is tricky (it's that 60-minutes-per-hour thing) but SPSS can work with it if you tell it what you're doing.

In the next example, we'll combine the separate hour and minutes fields into a new field that contains both. SPSS is good at transforming data this way. To build the new field, do the following:

1. **In the SPSS Data Editor window, choose Transform⇨Date and Time Wizard.**

   The window shown in Figure 3-11 appears.

2. **Select the option titled Create a Date/Time Variable from Variables Holding Parts of Dates or Times.**

3. **Click Next.**

4. **Put the names of the variables into the appropriate fields.**

   We want only the hours and minutes, so ignore the others. You move them by selecting the one you want from the list on the left, and then clicking the arrow next to the place you want it to go. When you're finished, the screen should look like Figure 3-12.

5. **Click Next.**

6. **Enter a name and a label for the variable. Also select a display format from the list.**

   To follow along with the example, type **time** in the Result Variable box, type **hour and minute** in the Variable Label box, and then select **hh:mm** in the Output Format list, as shown in Figure 3-13.

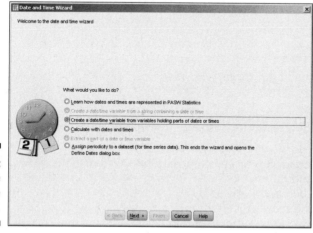

Figure 3-11:
The Date
and Time
Wizard.

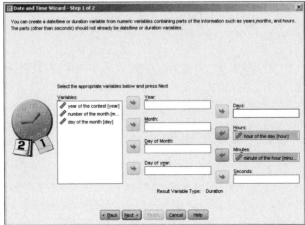

Figure 3-12:
Selection
of the
variables
from which
time is
structured.

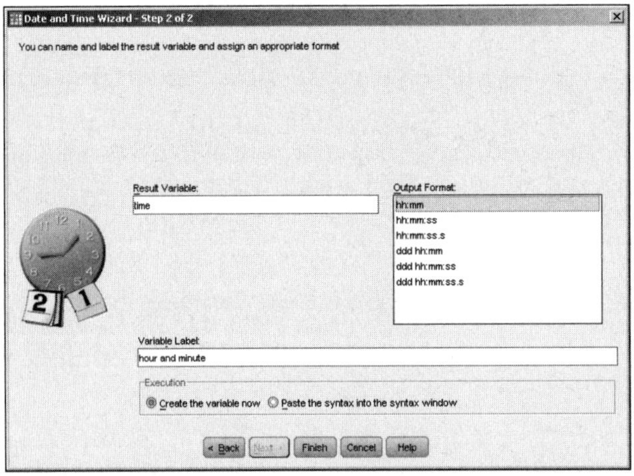

Figure 3-13:
The name
and display
format for
times.

7. **Select the Create the Variable Now option, and then click the Finish button.**

   You've created your new time data field. The result is shown in Figure 3-14.

Now follow the same procedure as before by choosing Analyze⇨Descriptive Statistics⇨Descriptives. But in Step 2, select only the new time field so you can see how SPSS handles different combinations of values. In the results, look at the difference in the two means: When the minutes are included, the mean moves to a time a bit later (as one would expect). It's now at 15:09 (3:09 PM) instead of 2-something. Whether that difference is statistically significant is up to you.

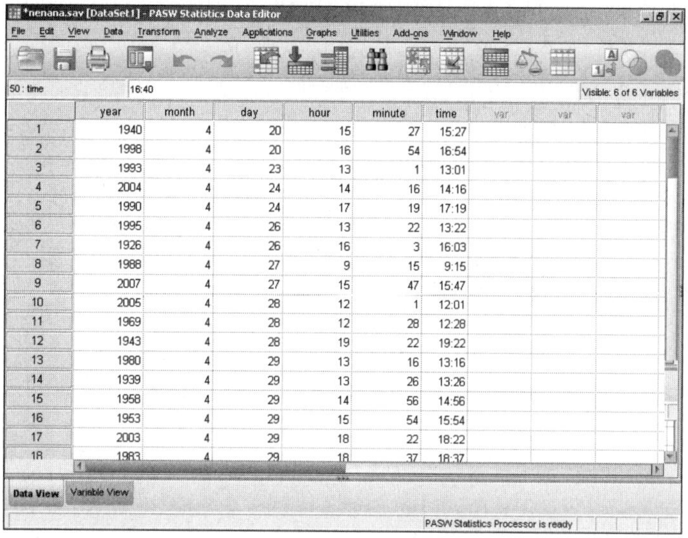

Figure 3-14:
The Data
Editor
window
with the
new time
field.

The Two Kinds of Numbers

With this example data so far, we have dealt with continuous variables.
Continuous variables are amounts and distances, such as age, gallons of gas,
and the number of beans in a jar. The other type is *categorical variables*. Here
you find things such as yes and no (where, for example, yes is 1 and no is 0)
and types of balls (where 1 is a football, 2 is a soccer ball, 3 is a snooker ball,
and so on). Each value represents a category.

All the variables in this example — except the number indicating the month —
are continuous variables. We tend to think of the months by their names instead
of numbers, but you have to use the number of the month to do any calculations.
If you want the name displayed, you have to assign a descriptive name for each
possible value. That's easy to do with this data because we have only two values:
4 and 5.

To add identifiers for the values, do the following:

1. **In the Data Editor window, click the Variable View tab and then
   select the cell in the Values column of the variable holding the month
   values.**

2. **Click the button that appears in the cell.**

3. **For each possible value, enter the value and the name you want associated with it, and then click Add.**

   The value, with its identifier, appears in the list, as shown in Figure 3-15.

4. **After you've added all the values you want to define, click OK.**

The screen displays only part of the change. The word None is gone and in its place is part of one of your new value definitions. But the real result will show up in your output and help you make a lot more sense of your results. For example:

1. **Choose Graphs⇨Legacy Dialogs⇨Pie.**

   The window shown in Figure 3-16 appears.

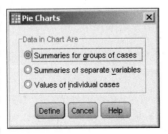

Figure 3-16:
Select the
type of
data to be
displayed
in the pie
chart.

2. **Select the Summaries for Groups of Cases option, and then click the Define button.**

   The window shown in Figure 3-17 appears.

3. **In the column on the left, select** number of the month, **then click the arrow to the left of Define Slices By, as shown in Figure 3-17.**

4. **Click the OK button.**

   SPSS Viewer appears, as shown in Figure 3-18.

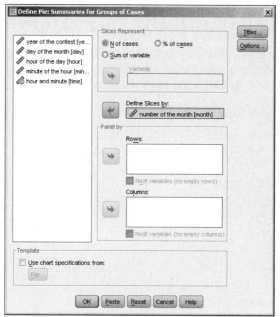

Figure 3-17:
You can
select the
variables
you want for
the pie
divisions.

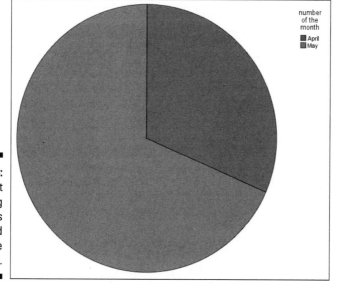

Figure 3-18:
A pie chart
including
the names
you defined
for the
values.

The Day It Is Most Likely to Happen

You already know that the ice is most likely to move in the warmer part of the day. A quick graph can show you whether or not there's a most likely day as well. To get a quick bar graph, do the following:

1. **Choose Graphs⇨Legacy Dialogs⇨Bar.**

   The dialog box shown in Figure 3-19 appears.

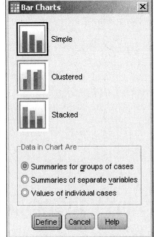

Figure 3-19:
You can select the fundamentals of the bar chart you want.

2. **Select the Simple bar chart and the Summaries for Groups of Cases option, and then click the Define button.**

3. **For Bars Represent, select N of cases, which means the bars will represent the number of cases. Also set the Category Axis to be the day of the month (day) and set the Rows to be the number of the month (month), as shown in Figure 3-20.**

   The exact meanings of these terms and settings are explained in Part III, which covers graphs.

4. **Click the OK button, and the bar chart in Figure 3-21 appears.**

The resulting chart shows which days in the past were most often the ones on which the ice moved. There is no obvious trend that I can see. However, you might want to experiment with different analysis displays and try to find a pattern.

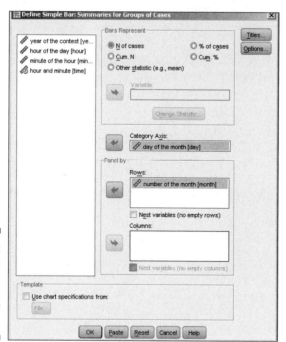

Figure 3-20:
Selecting
the data
to include
in the bar
chart.

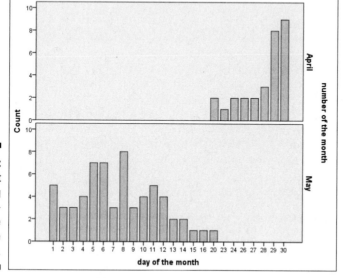

Figure 3-21:
A bar chart
showing
the distribu-
tion of the
days the ice
melts.

Part II
Getting Data In and Out of SPSS

The 5th Wave   By Rich Tennant

"We should cast a circle, invoke the elements, and direct the energy. If that doesn't work, we'll read the manual."

In this part . . .

The purpose of SPSS is to crunch numbers to come up with other numbers. To do the crunching, of course, you have to get the numbers into the program. After the crunching is finished, you have to get the numbers back out so you can see them. In fact, with the single exception of robotics, the sole purpose of every computer program in the world is to contain numbers and display them to a human.

Input can be tedious, but SPSS has some ways of helping ease the pain. Regular output is automatic, but you can do some special operations to handle irregular output. Either way, SPSS has you covered.

Chapter 4

Entering Data from the Keyboard

• •

In This Chapter

▶ Considering your choices when defining a variable

▶ Defining variables

▶ Entering numbers

▶ Making sure that you're using the right measurement type

• •

*T*o process your data, you have to get it into the computer. Entering data is always difficult; it has been a problem with computers since the beginning. No matter how you decide to get your numbers into SPSS, at some point someone has to type them (unless they come from some form of automatic monitoring). SPSS can read data from other places. You can also type directly into SPSS — and, if you want, copy the data to places other than SPSS later.

Entering data into SPSS is a two-step process: First you define what sort of data you will be entering, and then you enter the actual numbers. This may sound difficult, but it isn't so bad. When you see how data entry works in SPSS, you'll discover you have some pretty nifty software to help you.

You organize your data into cases. Each case is made up of a collection of variables. First, you define the characteristics of the variables that make up a case, and then you enter the data into the variables to make up the contents of the cases. This chapter shows you how to work with this technique of getting data into your system.

The Variable View Is for Entering Variable Definitions

You use the Variable View, shown in Figure 4-1, to define the names and characteristics of variables. This is where you always start if you plan on entering data into SPSS. You get to this window by clicking the Variable View tab at the bottom of the Data Editor window of SPSS. As you can see in Figure 4-1, every characteristic you can define about your variables is named at the top of the window. All you have to do is enter something in each column for each variable.

The predefined set of 12 characteristics are the only ones needed to completely specify all the attributes of any variable. The characteristics are all known to the internal SPSS processing. When you add a new variable, you will find that reasonable defaults appear for most characteristics.

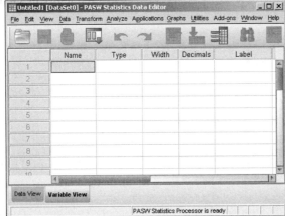

Figure 4-1:
You use
Variable
View to
define the
charac-
teristics of
variables.

The Variable View window is just for defining the variables. The entry of the actual numbers comes later. (See the section "The Data View Is for Entering and Viewing Data Items" later in this chapter.)

Each variable characteristic has a default, so if you don't specify a characteristic, SPSS fills one in for you. However, what it selects may not be what you want, so let's look at all the possibilities.

Name

The cell on the far left is where you enter the name of the variable. Just click in the cell and type a short descriptor such as age, income, sex, or odor. A longer descriptor, called a *label,* comes later. You could type longer names here, but you should keep them short because they'll be used in named lists and as identifier tags on the data graphs and such — where the format can be a bit crowded. Names that are too long can cause the output from SPSS to be garbled or truncated.

If the name you assigned turns out to be too long or is misspelled, you can always pop back into Variable View and change it. One of the nice things about SPSS is that you can correct mistakes quickly. I like that. (I had to hide a lot of them for the screen shots in this book.)

Here are some handy hints about names:

✔ You can use some bizarre characters in a name, such as @, #, and $, as well as the underscore character (_) and numbers.

If you decide you want to use some screwy characters in a name, go ahead and try it. It can't hurt. SPSS never does anything about it other than make you type something else.

✔ Be sure to start every name with an uppercase or lowercase letter.

✔ You can't include blanks anywhere in a name.

If you want to export data to another application, make sure the names you use are in a form acceptable to that application. Watch out for special characters.

Type

Most data you enter will be just regular numbers. Some, however, will be a special type, such as currency, and some will be displayed in a special format. Other data, such as dates, will require special procedures for calculation. You simply specify what type you have and SPSS takes care of those other details for you.

Click the cell in the Type column you want to fill in, and a button with three dots appears on its right. Click that button and the dialog box shown in Figure 4-2 appears.

Figure 4-2: The Variable Type dialog box allows you to specify the type of variable you are defining.

You can choose from the following predefined types of variables:

- ✔ **Numeric:** Standard numbers in any recognizable form. The values are entered and displayed in the standard form, with or without decimal points. Values can be formatted in standard scientific notation, with an embedded *E* to represent the start of the exponent. The Width value is the total number of all characters in a number — including any positive or negative signs and the exponent indicator. The Decimal Places value specifies the number of digits displayed to the right of the decimal point, not including the exponent.

- ✔ **Comma:** This type specifies numeric values with commas inserted between three-digit groups. The format includes a period as a decimal point. The Width value is the total width of the number, including all commas and the decimal point. The Decimal Places value specifies the number of digits to the right of the decimal point. You may enter data without the commas, but SPSS will insert them when it displays the value. Commas are never placed to the right of the decimal point.

- ✔ **Dot:** Same as Comma, except a period character (.) is used to group the digits into threes, and a comma is used for the decimal point.

- ✔ **Scientific Notation:** A numeric variable that always includes the *E* to designate the power-of-ten exponent. The *base*, the part of the number to the left of the *E*, may or may not contain a decimal point. The *exponent*, the part of the number to the right of the *E* — which also may or may not contain a decimal — indicates how many times 10 multiplies itself, after which it's multiplied by the base to produce the actual number. You may enter *D* or *E* to mark the exponent, but SPSS always displays the number using *E*. For example, the number 5,286 can be written as 5.286E3. To represent a small number, the exponent can be negative. For example, the number 0.0005 can be written as 5E–4. This format is useful for very large or very small numbers.

- ✔ **Date:** A variable that can include the year, month, day, hour, minute, and second. When you select Date, the available format choices appear in a list on the right side of the dialog box, as shown in Figure 4-3. Choose the format that best fits your data. Your selection determines how SPSS will format the contents of the variable for display. This format also determines, to some extent, the form in which you enter the data. You can enter the data using slashes, colons, spaces, or other characters. The rules are loose — if SPSS doesn't understand what you enter, it tells you, and you can re-enter it another way. For example, if you select a format with a two-digit year, SPSS accepts and displays the year that way, but it will use four digits to perform calculations. The first two digits (the number of the century) will be selected according to the configuration you set by choosing Edit⇨Options and then clicking the Data tab.

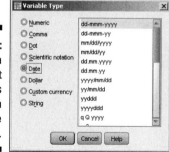

Figure 4-3:
Selecting a date format also selects which items are included.

✔ **Dollar:** When you select Dollar, the available format choices appear in a list on the right side of the dialog box. Dollar values are always displayed with a leading dollar sign and a period for a decimal point, and, for large values, will include commas to collect the digits in groups of threes. You select the format and its Width and Decimal Places values, as shown in Figure 4-4. The format choices are similar, but it's important that you choose one that's compatible with your other dollar-variable definitions so they line up when you print and display monetary values in output tables. The Width and Decimal Places settings help with vertical alignment in the output, no matter how many digits you include in the format itself. No matter what format you choose, you can enter the values without the dollar sign and the commas; SPSS inserts those for you.

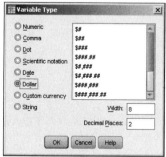

Figure 4-4:
The different dollar formats mostly specify the number of digits to be included.

✔ **Custom Currency:** The five custom formats for currency are named CCA, CCB, CCC, CCD, and CCE, as shown in Figure 4-5. You can view and modify the details of these formats by choosing Edit➪Options and then clicking the Currency tab. Fortunately, you can modify the definitions of these custom formats as often as you like without fear of damaging your data. As with the Dollar format, the Width and Decimal Places settings are primarily for aligning the data when you're printing a report.

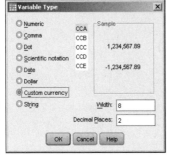

Figure 4-5:
Five custom currency formats are available.

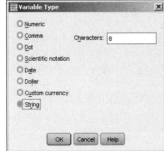

✔ **String:** A freeform non-numeric item. Because it is non-numeric, the contents of a variable of this type can never be used for calculations. You can specify any number of any characters up to the maximum length you specify, as shown in Figure 4-6. You can also use a variable of this type as a descriptor or an identifier of a particular case.

Figure 4-6:
A freeform type never used in calculations.

Width

The width setting in the definition of a variable determines the number of characters used to display the value. If the value to be displayed is not large enough to fill the space, the output will be padded with blanks. If it is larger than you specify, it will either be reformatted to fit or asterisks will be displayed.

Certain type definitions allow you to set a width value. The width value you enter as the Width definition is the same as the one you enter when you define the type. If you make a change to the value in one place, SPSS changes the value in the other place automatically. The two values are the same.

At this point, you can do one of three things:

- ✔ Skip this cell and accept the default (or the number you entered previously under Type).
- ✔ Enter a number and move on.
- ✔ Use the up and down arrows that appear in the cell to select a numeric value.

Decimals

The number of decimals is the number of digits that appear to the right of the decimal point when the value appears on-screen. This is the same number that you may have specified as the Decimal Places value when you defined the variable type. If you entered a number there, it appears here as the default. If you enter a number here, it changes the one you entered for the type. They are the same.

Now you can do one of three things:

- ✔ Skip this cell and accept the default (or the number you entered earlier under Type).
- ✔ Enter a number and move on.
- ✔ Use the up and down arrows that appear in the cell to select a numeric value.

Label

The name and the label serve the same basic purpose: They are descriptors that identify the variable. The difference is that the *name* is the short identifier and the *label* is the long one. You need one of each because some output formats work fine with a long identifier and other formats need the short form.

You can use just about anything for the label. What you choose has to do with how you expect to use your data and what you want your output to look like. For example, the name may be sex and the longer label may be Boys and Girls, Men and Women, or simply Gender.

The length of the label is not determined by some sort of software requirement. However, output looks better if you use short names and somewhat longer labels. Each one should make sense standing alone. After you produce some output, you may find that your label is lousy for your purposes. That's okay; it's easy to change. Just pop back to the Variable View and make the change. The next time you produce output, the new label will be used.

You can also just skip defining a label. If you don't have a label defined for a variable, SPSS will use the name you defined for everything.

Value

The Values column is where you assign labels to all the possible values of a variable. If you select a cell in the Values column, a button with three dots appears. Clicking that button displays the dialog box shown in Figure 4-7.

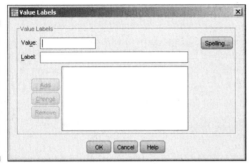

Figure 4-7:
You can assign a name to each possible value of a variable.

Normally, you make one entry for each possible value that a variable can assume. For example, for a variable named Sex you could have the value 1 assigned the label Male and 2 assigned Female. Or, for a variable named Committed you could have 0 for No, 1 for Yes, and 2 for Undecided. If you have labels defined, when SPSS displays output, it will show the labels instead of the values.

To define a label for a value:

1. **In the Value box, enter the value.**

2. **In the Label box, enter a label.**

3. **Click the Add button.**

   The value and label appear in the large text block. To change or remove a definition, simply select it in the text box and make your changes.

4. **Repeat Steps 1–3 as needed.**

5. **Click the OK button to save the value labels and close the window.**

You can always come back and change the definitions, using the same process you used to enter them. The window will reappear, filled in with all the definitions; then you can update the list.

Missing

You can specify what is to be entered for value that is missing for a variable in a case. That is, when you have values for all variables in a case except one, you can specify a placeholder for the missing value. Click a cell in the Missing column, and the dialog box shown in Figure 4-8 appears.

For example, say you are entering responses to questions, and one of the questions is, "How many dirigibles do you own?" The normal answer to this question is a number, so you define the variable type as a number. If someone chooses to ignore this question, this variable won't have a value. However, you can specify a placeholder value. Perhaps 0 seems like a good choice for a placeholder here, but it's not really: A common answer will be 0. Instead, a less likely value — like, say, –1 — makes a better choice.

You can even specify unique values to represent different reasons for a value being missing. In the previous example, you could define –1 as the value entered when the answer is, "I don't remember," and –2 could be used when the answer is, "None of your business." If you specify that a value is representing a missing value, that value is not included in general calculations. During your analysis, however, you can determine how many values are missing for each of the different reasons. You can specify up to three specific values (called *discrete values*) to represent missing data, or you can specify a range of numbers along with one discrete value, all to be considered missing. The only reason you would need to specify a range of values is if you have lots of reasons why data is missing and want to track them all.

Columns

Columns is where you specify the width of the column you will use to enter the data. The folks at SPSS could have used the word *Width* to describe it, but they already used that term for the width of the data itself. A better name might have been the two words *Column Width,* but that would have been too long to display nicely in this window, so they just called it *Columns.* To specify the number of columns, select a cell and enter the number.

Align

The Align column determines the position of the data in its allocated space, whenever the data is displayed for input or output. The data can be left-aligned, right-aligned, or centered. You've defined the width of the data and the size of the column in which the data will be displayed; the alignment determines what is done with any space left over.

When you select a cell in the Align column, a list appears and you can choose one of the three alignment possibilities, as shown in Figure 4-9. Aligning to the left means inserting all blanks on the right; aligning right inserts all the extra spaces on the left; centering the data splits the spaces evenly on each side — but I don't know what it does if an odd space is left over. (I also worry about things like the number of seeds in a tomato and where the clouds go at night.)

Figure 4-9:
Values can
be justified
right or left,
or posi-
tioned in the
center.

Measure

Your value here specifies the measure of something in one of three ways. When you click a cell in the Measure column, you can select one of these choices (see Figure 4-10):

 ✔ **Scale:** A number that specifies a magnitude. It can be distance, weight, age, or a count of something. Most numbers fall into this category. The technical name for this type of number is *cardinal,* but SPSS uses Scale to keep life simple.

✔ **Ordinal:** These numbers specify the position (order) of something in a list. For example, *first*, *second*, and *third* are ordinal numbers.

✔ **Nominal:** Numbers that specify categories or types of things. You can have 0 represent Disapprove and 1 represent Approve. Or you can use 1 to mean Fast and 2 to mean Slow.

Figure 4-10:
The type of measure-ment being made by the values in this variable.

Role

Some of the SPSS dialog boxes select variables according to their role and include them as defaults. You don't need to worry about this characteristic. It can be handy after you have some experience with SPSS and understand how defaults are chosen. When you click a cell in the Role column, you can select one of six choices (see Figure 4-11):

Figure 4-11:
The role assumed by this variable in certain SPSS dialog boxes.

✔ **Input:** This variable is used for input. This is the default role. Definition of Roles is new to version 18 of SPSS, and all data imported from earlier versions will be assigned this role.

✔ **Target:** This variable is used as output by SPSS procedures.

✔ **Both:** This variable is used as both input and output.

✔ **None:** This variable has no role assignment.

✔ **Partition:** This variable is used to partition the data into separate samples for training, testing, and validation.

✔ **Split:** This option is included for round-trip compatibility with the SPSS modeler. This capability, however, should not be confused with file splitting (described in Chapter 16).

The Data View Is for Entering and Viewing Data Items

After you've defined all the variables for each case, switch the display to the Data View so you can begin typing the data. You make the switch by clicking the Data View tab at the bottom of the window. When you do, the Data Editor window appears.

At the top of the columns in Figure 4-12, you can see some names I chose for variables. Switching to Data View makes the window ready to receive entered data — and to verify that what's entered matches the specified format and type of the data.

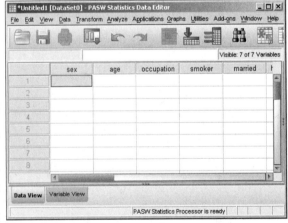

Figure 4-12:
The Data Editor window, ready to accept new data.

Entering data into one of these cells is straightforward: You simply click the cell and start typing.

If something is already in a cell and you want to change it instead of just typing over it, look up toward the top of the window, just underneath the toolbar: You'll see the name of the variable and the currently selected value. Click the value in the field at the top, and you can edit it right there. You can do all the normal mouse and keyboard stuff there, too — you can use the Backspace key to erase characters, or select the entire value and type right over it.

If you feel like a lousy (or inexperienced) mouse driver, take some time to experiment and figure out how to edit data. Lots of software use these same editing techniques, so becoming proficient now will pay you dividends later.

If your data is already in a file, you might be able to avoid typing it in again by reading that file directly into SPSS. For more information, see Chapter 5.

Don't take chances. As soon as you type a few values, save your data to a file by choosing File➪Save As. Then choose File➪Save throughout the process of entering data, and you won't be ruined when the computer crashes unexpectedly.

We all have to go back and refine our variable definitions from time to time. That's normal. When you come across something that doesn't do what you want it to, just switch back to Variable View and correct it. Nobody but you and SPSS will ever know about it, and SPSS never talks.

Filling In Missed Categorical Values

Now that you have defined your variables and entered your data, you might want to check that you have names defined for all your actual ordinal and nominal values, and that you have defined the correct measures for them. SPSS can help by scanning your data, finding values for which you don't have definitions, and pointing them out in a friendly way.

The following steps use an existing file to walk through a demonstration:

1. **Choose File➪Open➪Data to load the file named** Cars.sav.

   This file came with your installation of SPSS and is found, along with a number of other files, in the same directory in which you installed SPSS. You can load any of these data files, but Cars.sav is the one used in this demonstration. If you load this file while you already have some other data showing in the window, SPSS will open a new Data Editor window to display the new information; your existing data will not be lost.

When you open this data file — or any data file, for that matter — SPSS opens a SPSS Viewer window to tell you that it has opened a file (or the information could be displayed in a SPSS Viewer window that is already open). You won't need this information for what you are doing here, so you can just close the window.

2. **Choose Data⇨Define Variable Properties.**

   The Define Variable Properties dialog box appears.

3. **On the left, select all the names of the variables you want to check, and then click the arrow in the center of the window to move them to the right, as shown in Figure 4-13.**

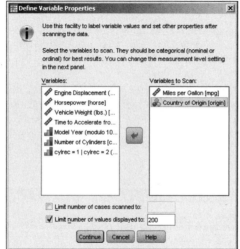

Figure 4-13:
Selecting
variables to
check their
properties.

4. **Click the Continue button.**

5. **Select one of the variable names in the list on the left.**

   Its different values appear in the center of the window, as shown in Figure 4-14. (In this example, every value has a name assigned to it.)

6. **Ask SPSS to suggest a new type for this variable by clicking the Suggest button in the top center of the window**

   The window in Figure 4-15 appears, telling you what SPSS concludes about this variable and its values. This same window, with different text, appears for each variable you test. Sometimes the text suggests changes in the variable definition, and sometimes it does not.

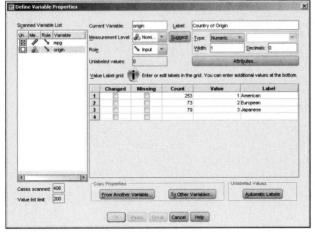

Figure 4-14:
The values of the selected variable.

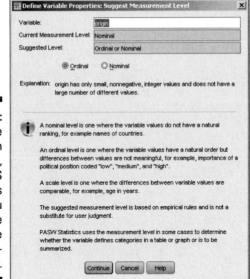

Figure 4-15:
From the pattern of values, SPSS concludes whether you may have chosen the wrong measurement.

7. To apply any changes, click Continue.

You return to the window shown in Figure 4-14, where you can select another variable.

You won't want to make changes to all your variables, but SPSS helps you find the ones that you do need to change. Values defined as Missing are not included in the computations. The text in the window always explains the criteria used to reach a conclusion, and SPSS allows you to make the final decision.

Chapter 5

Reading and Writing Files

. . . .

In This Chapter

▶ Understanding the SPSS file format

▶ Reading a simple text file into SPSS

▶ Transferring data from another application into SPSS

▶ Saving SPSS data in various formats

. . . .

There is no need to put your data into the computer more than once. If you've entered your data in another program, you can copy it from there into SPSS — because every program worth using has some form of output that can serve as input to SPSS. This chapter discusses ways to transfer data into and out of SPSS.

The SPSS File Format

SPSS has its own format for storing data and writes such files with the .sav extension. This file format contains special codes and usually can't be used to export your data to another application. It is used only for saving SPSS data that you want to read back into SPSS at a later time. Several example files in this format are copied to your computer as part of the normal SPSS installation. These files can be found in the same directory as your SPSS installation. You can load any one of them by choosing File⇨Open⇨Data and selecting the file to be loaded. When you do so, the variable names and data are loaded and fill your SPSS window.

If you have SPSS filled with data, you can save it to a .sav file by choosing File⇨Save As and providing a name for the file. Or if you've loaded the information from a file, or have previously saved a copy of the information to a file, you can simply use the File⇨Save selection to overwrite the previous file with a fresh copy of both variable definitions and data.

You can be fooled by the way SPSS help uses the word *file*. If you have defined data and variables in your program, the SPSS documentation often refers to it all as a file, even though it may have never been written to disk. They also refer to the material written to disk as a file, so watch the context.

When you write your file to disk, if you don't add the .sav (or .SAV) extension to the filename, SPSS adds it for you. When you use File⇨Open⇨Data to display the list of files, you may or may not see the extension on the filename (it depends on how your Windows system is configured), but it's there.

Formatting a Text File for Input into SPSS

If your data is in an application that can't directly create a file of a type that SPSS can read, getting the data into SPSS may be easier than you think. If you can get the information out of your application and into a text file, however, it's fairly easy to have SPSS read the text file.

When it comes to writing information to disk, some applications are more obliging than others. Look for an Export menu option — it usually has some options that allow you to organize the output text in a form you want. (Read on for a description of possible organization schemes.)

If the application doesn't allow you to format text the way you want, look for printer options — maybe you can redirect printer output to a disk file and work from there. If you use the application's printer output, you may need to use your word processor to clean up the form of the data. I know this multi-step operation sounds like a lot of work, but it's often easier than typing all your data in again by hand.

The data file you output from SPSS doesn't have to include the variable names, just the values that go into the variables. You can format the data in the file by using spaces, tabs, commas, or semicolons to separate data items. Such dividers are known as *delimiters*. Another method of formatting data avoids delimiters altogether. In that method, you don't have to separate the individual data items, but you must make each data item a specific length, because you have to tell SPSS exactly how long each one is.

The most intuitive format is to have one case (one row of data) per line of text. That means the data items in your text file are in the same positions they will be in when they are read into SPSS. Alternatively, you can have all your data formatted as one long stream, but you'll have to tell SPSS how many items go into each case.

Always save this kind of raw data as simple text; the file you store it in should have the `.txt` (or `.TXT`) extension so SPSS can recognize it for what it is.

Reading Simple Data from a Text File

This section contains an example of a procedure you can follow to read data from a simple text file into SPSS. The file is a simple file named `garbler.txt`. It contains two cases (rows of data) as two lines of text, with the data items in the two lines separated by spaces. The content of the file is as follows:

```
"Pat" 1 35 3.00 9
"Chris" 1 22 2.4 7
```

The following example reads this text file and inserts it into the cells of SPSS. Along the way, SPSS keeps you informed about what's going on so there won't be any big surprises at the end.

1. **Choose File➪Read Text Data.**

   The file selection window shown in Figure 5-1 appears.

Figure 5-1:
Locate the file you want to read.

2. **Select the `garbler.txt` file, and then click the Open button.**

   The screen shown in Figure 5-2 appears, for loading and formatting your data.

3. **Examine the input data.**

   The screen lets you peek at the contents of the input file so you can verify that you've chosen the right file. Also, if your file uses a predefined format (which it doesn't, in this example), you can select it here and skip some of the later steps. If your data doesn't show up nicely separated into values the way you want, you may be able to correct it in a later step. Don't panic just yet.

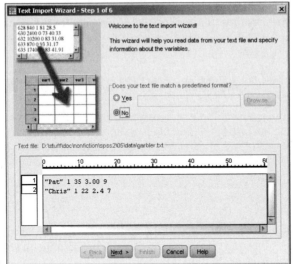

Figure 5-2:
Make cer-
tain your
data looks
reasonable.

4. **Click the Next button.**

The screen shown in Figure 5-3 appears.

Figure 5-3:
Specify
whether the
fields are
delimited
and whether
the variable
names are
included.

5. **Specify that the data is delimited and the names are not included.**

As you can see in this example, SPSS takes a guess, but you can also specify how your data is organized. It can be divided using spaces (as in this example), commas, tabs, semicolons, or some combination. Or your data may not be divided — it may be that all the data items are jammed together and each has a fixed width. If your text file includes the names of the variables (I'll show you how this works in a minute), you need to tell SPSS.

6. Click the Next button.

The screen shown in Figure 5-4 appears.

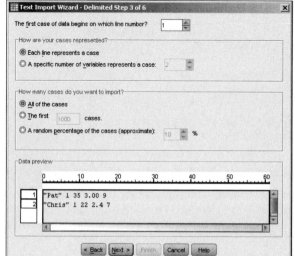

Figure 5-4:
Specify
where
the data
appears in
the file.

7. Specify how SPSS is to interpret the text.

For this example, the correct settings are shown in Figure 5-4. You can tell SPSS something about the file and which data you want to read:

- Perhaps some lines at the top of the file should be ignored — this happens when you're reading data from text intended for printing and header information is at the top. By telling SPSS about it, those first lines can be skipped.

- Also, you can have one line of text represent one case (one row of data in SPSS), or you can have SPSS count the variables to determine where each row starts.

- And you don't have to read the entire file — you can select a maximum number of lines to read starting at the beginning of the file, or you can select a percentage of the total and have lines of text randomly selected throughout the file. Specifying a limited selection can be useful if you have a large file and would like to test parts of it.

8. Click the Next button.

The screen shown in Figure 5-5 appears.

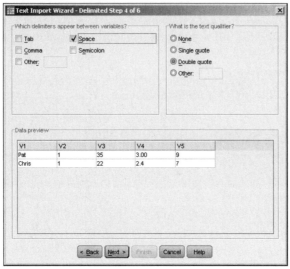

Figure 5-5:
Specify the
delimiters
that go
between
data items
and which
quotes to
use for
strings.

9. **Specify space as the delimiters and double quotes as text qualifiers.**

   SPSS knows how to use commas, spaces, tabs, and semicolons as delimiting characters. You can even use some other character as a delimiter by selecting Other and then typing the character into the blank. You can also specify whether your text is formatted with quotes (as in our example) and whether you use single or double quotes. Strings must be surrounded in quotes if they contain any of the characters being used as delimiters.

   You can specify that a data item is missing in your text file. Simply use two delimiters in a row, without intervening data.

10. **Click the Next button.**

    The screen shown in Figure 5-6 appears.

11. **Change the variable names and types, if you wish.**

    SPSS assigns the variables the names V1, V2, V3, and so on. To change a name, select it in the column heading at the bottom of the window, and then type the new name in the Variable Name field at the top. You can select the format from the Data Format pull-down list, as shown in Figure 5-6. This is optional. If you need to refine your data types and whatnot, you can do so later in the Variable View window. The point here is to get the data into SPSS.

12. **Click the Next button.**

    The screen shown in Figure 5-7 appears.

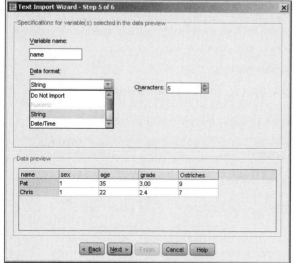

Figure 5-6:
Name your
variables
and select
their data
types.

Figure 5-7:
Save the
format, grab
the syntax,
or enable
caching.

13. Choose "No" to not save this file format.

This is something you would do if you'll be loading more files of this
same format into SPSS — it reduces the number of questions to answer
and the amount of formatting to do next time. You also have the chance
to grab a copy of the Syntax Language instructions that do all this, but
unless you know about the Syntax Language (as described in Chapters
15 and 16), it's best to pretend that option doesn't exist. (For that

matter, the Cache Data Locally option is a bit odd. I don't know why it's there, unless SPSS has some problem with huge files. SPSS seems to load data faster with it than without it, but it's strictly an internal thing and SPSS works just fine either way.)

14. Click the Finish button.

Depending on the type of data conversions and the amount of formatting, SPSS may take a bit of time to finish. But be patient; the SPSS Data View window will eventually display your data.

15. Look at the data. Correct your data types and formats, if necessary. Then save it all to a file by choosing File⇨Save As.

You are instructed to enter a filename. You can just call it `garbler`. The new file will have the `.sav` extension, which indicates that it's a standard SPSS file.

The SPSS way of reading data is a lot more flexible than this simple example demonstrates. Another example can help show why. Here, a file named `headgarbler.txt` is that same data, formatted slightly differently:

```
Name Sex Age GradePoint Ostriches
Pat,1,35,3.00,9,Chris,1,22,2.4,7
```

This time the data in the file is preceded by the variable names listed on the first line, the data is all in one long line, and the data is separated by commas. To read this into SPSS, you start the same way you did before. However, SPSS can't figure it all out in Step 1 this time (as shown in Figure 5-8). SPSS can't even tell which is header and which is data.

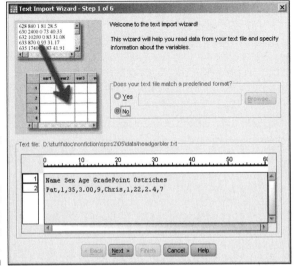

Figure 5-8:
The data remains as a block of text until you explain the parts.

In Step 2 of 6, you select the option that informs SPSS that the variable names appear in the first line of text. Then, in Step 3 of 6 (as shown in Figure 5-9), you specify that the data begins on line 2 of the text file. It's possible for the data to begin several lines down in the input text file, but if variable names are present, they must be on the first line. Also, when you specify variable names, SPSS ignores the beginning and ending of lines, and counts the data values to determine when it has a complete row (case).

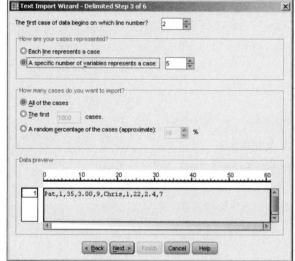

Figure 5-9:
Specify that
the data
starts on
line 2 and
each case
has 5 data
items.

In Step 4 of 6 (shown in Figure 5-10), commas and spaces were chosen as delimiters. (Although no spaces appear in the data in this example, it doesn't hurt to include a space delimiter if it may occur somewhere in your data.) Also, None was chosen for the characters surrounding string values. In this example, SPSS figured the spacing out on its own and used these settings for its default. Also, by the time you reach Step 4 of 6, SPSS has started organizing the data according to your definitions. It has already read the variable names and included them as column headers.

In Step 5 of 6, you have the opportunity to change the variable names and specify their types. Here again, you see that SPSS has made a guess for the type of each one.

After you complete Step 6 of 6, click the Finish button and wait for the data to load, as shown in Figure 5-11.

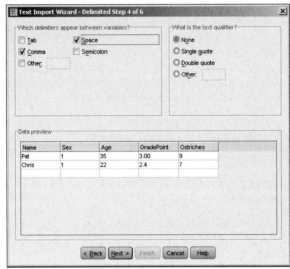

Figure 5-10:
Specifying
delimiters
and quote
characters.

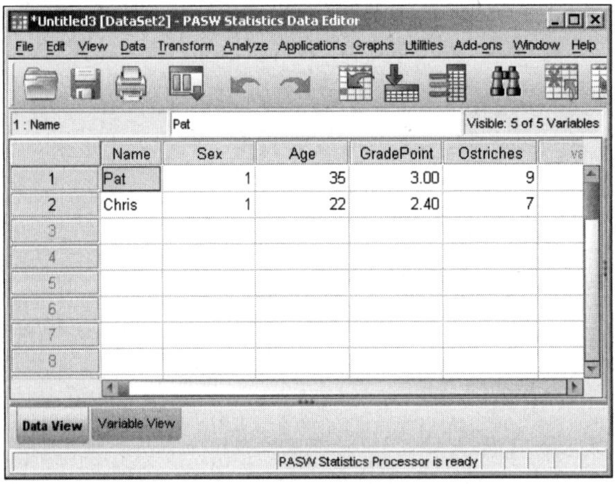

Figure 5-11:
The data as
formatted in
SPSS.

You can see who has how many ostriches, but you still have a little work
to do. For example, switch to Variable View, change the sex variable to
a nominal data type, and assign the names "male" and "female" to the
values 1 and 2. (You can't assume anything about sex by the names.) You
might want to add some descriptive labels. For example, the variable named
"ostriches" could be given the descriptive name "ostrich count in
front yard". See how a good descriptive name can clear up a little
mystery?

Transferring Data from Another Program

You can get your data into SPSS from a file created by another program, but it isn't always easy. SPSS knows how to read some file formats, but if you're not careful you'll find your data stored in an odd file format, and deciphering some file formats can be as confusing as Klingon trigonometry. SPSS can read only from file formats it knows.

SPSS recognizes the file formats of several applications. Following is a complete list:

✔ **IBM SPSS Statistics (**. sav**):** IBM SPSS Statistics data, and also the format used by the DOS program SPSS/PC+.

✔ **dBase (**. dbf**):** An interactive database system.

✔ **Excel (**. xls**):** Spreadsheet for performing calculations on numbers in a grid.

✔ **Portable (**. por**):** A portable format read and written by other versions of SPSS, including other operating systems.

✔ **Lotus (**. w**):** Spreadsheet for performing calculations with numbers in a grid.

✔ **SAS (**. sas7bdat, . sdy, . sd2, . ssd**, and** . xpt**):** Statistical analysis software.

✔ **Stata (**. dta**):** Statistical analysis and graphics software.

✔ **Sylk (**. slk**):** A symbolic link file format for transporting data from one application to another.

✔ **Systat (**. syd **and** . sys**):** Software that produces statistical and graphical results.

Although SPSS knows how to read any of these, you may still need to make a decision from time to time about how SPSS should import your data set. But you have some advantages: You know exactly what you want (the form of data appearing in SPSS is simple, and what you see is what you get), SPSS has some reasonable defaults and makes some good guesses along the way, and also you can always fiddle with things after you've loaded them.

You are only reading from the data file, so you can't hurt it. Besides, you have everything safely backed up, don't you? Just go for it. If the process gets hopelessly balled up, you can always call it quits and start over. That's the way I do it — I think of it as my learning process.

Reading an Excel file

Here's an example. SPSS knows how to read Excel files directly. If you want to read the data from an Excel file, I suggest you read the steps in "Reading Simple Data from a Text File," earlier in this chapter, because the two processes are similar. If you understand the decisions you have to make in reading a text file, reading from an Excel file will be duck soup. Figure 5-12 shows the appearance of data displayed by Excel.

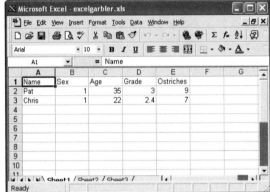

Figure 5-12:
A simple
example
of Excel
spreadsheet
data.

Do the following to read this data into SPSS:

1. **Save the Excel data to a file.**

   In this example, the file is called `excelgarbler.xls`. If you want to copy only a portion of the spreadsheet, make a note of the cell numbers in the upper-left and lower-right corners of the group you want.

2. **Close Excel.**

   You must stop the Excel program from running before you can access the file from SPSS.

3. **Choose File⇨Open⇨Data.**

4. **Select the `.xls` file type, as shown in Figure 5-13, and then click Open.**

5. **Select the data to include.**

   An Excel file can contain more than one sheet, and you can choose the one you want from the pull-down list, as shown in Figure 5-14. Also, if you've elected to read only part of the data, enter the Excel cell numbers of the upper-left and lower-right corners here. You specify the range of cells the same way you would in Excel — using two cell numbers separated by a colon. Don't worry about the maximum length for strings.

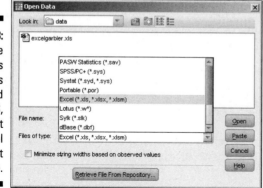

Figure 5-13:
From the
many types
of files
understood
by SPSS,
select
the Excel
spreadsheet
type.

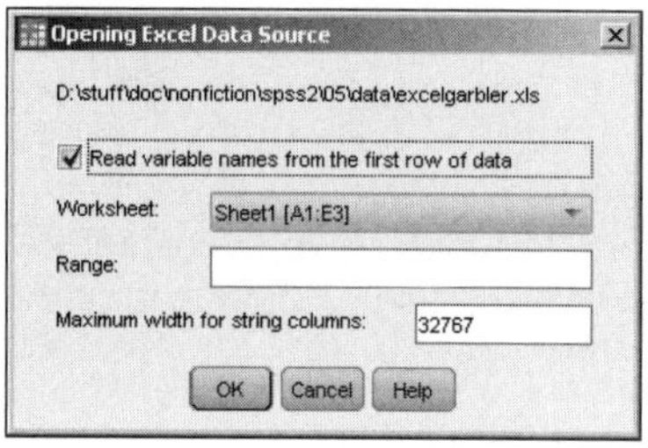

Figure 5-14:
Select
which
data in the
spreadsheet
to include.

6. **Click OK.**

   Your data appears in the SPSS window.

7. **Check your variables and adjust their definitions as necessary.**

   SPSS makes a bunch of assumptions about your data, and it probably makes some wrong ones. Closely examine and adjust your variable definitions by switching to Variable View and making the necessary changes.

8. **Save the file using your chosen SPSS name, and you're off and running.**

Reading from an unknown program type

Often you can transfer data from another application into SPSS by selecting, copying, and pasting the data you want, but that method has its drawbacks. The places you're copying from and to are usually larger than the screen, so highlighting and selecting can be tricky. You must be ready to choose Edit➪Undo when necessary.

A better method is to write the data to a file in a format understood by SPSS, and then read that file into SPSS. SPSS knows how to read some file formats directly. Using such a file as an intermediary means you have an extra backup copy of your data, and that's never a bad idea.

Saving Data and Images

Writing data from SPSS is easier than reading data into SPSS. All you do is choose File⇨Save As, select your file type, and then enter a filename. You have lots of file types to choose from. You can write your data not only in two plain-text formats, but also in Excel spreadsheet format, three Lotus formats, three dBase formats, six SAS formats, and six Stata formats.

If you'll be exporting data from SPSS into another application, find out what kinds of files the other application can read, and then use SPSS to write in one of those formats.

A second form of output from SPSS is an image. If you've generated a graphic that you want to insert into your word processor or place on your Web site, SPSS is ready to help you do it. (I almost wish it were hard to do so I could look smart showing you how, but it's easy.)

When you go through the steps to produce a graph, as explained in Part III, you'll be looking at the resulting graphics in the SPSS Viewer, which is shown in Figure 5-15.

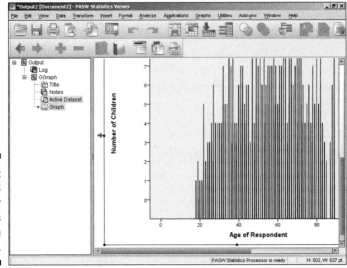

Figure 5-15:
SPSS
Viewer
displays
graphs on
the screen.

From SPSS Viewer, you can export images (and do some other things too):

1. **Produce a graph or table.**

   You can use any of the examples in Part III to produce a graphic display. SPSS Viewer pops up and displays the output.

2. **Choose File⇨Export.**

   The window shown in Figure 5-16 appears.

3. **In the Objects to Export section, select which items to include in the output.**

   You can elect to have all objects output, all visible objects output, or only the ones you've selected. In Figure 5-15, for example, the panel on the left indicates that two items — Active Dataset and Graph — were selected. The *visibility* of an object refers to whether its name appears in the list — if you collapse the list so a particular name can't be seen, the item is not visible. You can select items by clicking the items themselves, or by selecting their names in the list on the left.

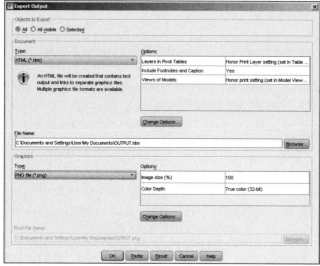

Figure 5-16: These selections control what gets output and into what format.

4. **In the Document section, open the Type pull-down list and choose an output format.**

   Your choices vary according to what you decided to output as specified at the top of the window. Here is a list of the possible file formats:

   - *Excel files* can include text, tables, and graphics, with the graphics embedded in the workbook. The data can create a new file or be added to an existing workbook. No graphic options are available.

- *HTML files* can be used for text both with and without graphics. If graphics are included, those will be exported separately, and they will be included as HTML links. The graphic file type must also be chosen.

- *PDF* documents exported will include not only text but also any graphics existing in the original. No graphics options are available.

- *PowerPoint documents* can be written as text with the graphics embedded in the TIFF format. No graphic options are available.

- *Text files* can be output with graphic references included, and the graphics written to separate files. The reference is the name of the graphic file. The graphic file format is specified by choosing options in the lower section of this window.

- *UTF-8* is Unicode text encoded as a stream of 8-bit characters. Graphics are handled the same as they are for text files.

- *UTF-16* is Unicode text encoded as a stream of 16-bit characters. Graphics are handled the same as they are for text files.

- *RTF:* Word documents are written in RTF (rich text format), which can be copied into a Word document. No graphic options are available.

- *None:* When selected, this option means no text is output — only graphic images. The graphic file format is specified by options in the lower section of this window.

5. **In the Graphics section, select the image file format, if one is needed, from the Type drop-down list.**

   You may be asked to select a format for your image file(s). You can select from PNG (.png), bitmap (.bmp), enhanced metafile (.emf), encapsulated postscript (.eps), jpeg (.jpg), or tagged image file (.tif).

6. **Select the directory and root filename, and click Save.**

   Click the Browse button, and you can select the directory and the root name of the file(s) you want to create. Depending on what you chose to output, the actual output may be multiple files, and they will all have names derived from the root name you provide. The Save button does not write the file(s) — it only inserts your selected name into the Export Output window.

7. **Click the OK button.**

   Doing so writes the file (or files) to disk — each in the chosen format, at the chosen location.

Chapter 6

Data and Data Types

● ●

In This Chapter

▶ Understanding the special properties of dates and times

▶ Working with data that comes at regular intervals

▶ Creating multiple response sets

▶ Copying variable definitions from another file

● ●

A data type is nothing more than the definition of what a number means. Without a definition, a number serves no purpose. For example, the number 3 could have entirely different meanings. It could be a number of miles, or an answer to a multiple-choice question, or the number of jelly beans in your left pocket. The data type is more than just a tag — it determines how the value can be manipulated. For example, 3 miles can also be written as 15,840 feet or as 24 furlongs. Some data types require special arithmetic. Telling time is an example: If the number 50 represents the number of minutes past 2 o'clock, adding 15 to it will result in the number 5 — that is, the number of minutes past *3* o'clock.

Dates, times, and schedules are important in statistics, but they're usually hard to work with arithmetically. Fortunately, all you have to do is tell SPSS how you'd like to handle them, and all the hard calculating can be taken care of for you. Arithmetic that normally would be tedious and boring can be automated by assigning the appropriate data types. And if you've worked on a table and arrived at some nifty variable definitions, you can copy them into a new table (or even into an old table). This chapter discusses the nuts and bolts of creating such magic by working with data and data types.

Dates and Times

Calendar and clock arithmetic can be tricky, but SPSS can handle it all for you. Just enter the date and time in whatever format you specify, and SPSS converts those values into its internal form to do the calculations. Also, SPSS displays the date and time in your specified format, so it's easy to read.

SPSS understands the meaning of slashes, commas, colons, blanks, and names in the dates and times you enter, so you can write the date and time almost any way you'd like. If SPSS can't figure out what you've typed, it clears away what you typed and waits for you to type something again.

Internally, SPSS keeps all dates as a positive or negative count of the number of seconds from a zero date. As a result, all dates also include the time of day. You can choose a display format that includes or excludes the time, but the information is always there. You can even change the display format without loss of data. If the time is not included in the data you enter, SPSS assumes zero hours and minutes (midnight).

You determine the data type for each variable in the Data View window. The type is chosen from the list of types shown in Figure 6-1. On the right, you select a format. SPSS uses this format to interpret your input and to format the dates for display.

Figure 6-1:
Select the
data type
and the
format.

SPSS uses the format you select for both reading your input and formatting the output of dates and times.

The Columns setting of the date variable in the Variable View is important. The column width determines the maximum number of characters that can be displayed, and if you choose a format that is too wide to fit, the date will show up only as a row of asterisks.

The available formats are defined as a group and change according to the variable type. For example, the Dollar type has a different list of choices from those offered for the Date type (shown in Figure 6-1).

The list of format definitions you have to choose from are constructed by combining the specifiers listed in Table 6-1. Format definitions look like mm/dd/yy and ddd:hh:mm.

Table 6-1	Specifiers in Date and Time Formats
Specifier	**Means**
dd	A two-digit day of the month in the range 01, 02, . . . , 30, 31.
ddd	A three-digit day of the year in the range 001, 002, . . . , 364, 365.
hh	A two-digit hour of the day in the range 00, 01, . . . , 22, 23.
Jan, Feb, . . .	The abbreviated name of the month of the year, as in JAN, FEB, . . . , NOV, DEC.
January, February, . . .	The name of the month of the year, as in JANUARY, FEBRUARY, . . . , NOVEMBER, DECEMBER.
mm	When adjacent to a dd specifier in a format, a two-digit month of the year in the range 01, 02, . . . , 11, 12. When adjacent to an hh specifier in a format, a two-digit specifier of the minute in the range 00, 01, . . . , 58, 59.
mmm	A three-character name of a month, as in JAN, FEB, . . . , NOV, DEC.
Mon, Tue, . . .	The abbreviated name of the day of the week, as in MON, TUE, . . . , SAT, SUN.
Monday, Tuesday, . . .	The name of the day of the week, as in MONDAY, TUESDAY, . . . , SATURDAY, SUNDAY.
q Q	The quarter of the year, as in 1 Q, 2 Q, 3 Q, or 4 Q.
Ss	Following a colon, the number of seconds in the range 00, 01, . . . , 58, 59. Following a period, the number of hundredths of a second.
ww WK	The one- or two-digit number of the week of the year in the range 1 WK, 2 WK, . . . , 51 WK, 52 WK. *Note:* Although week numbers can be either one or two digits, the numbers always line up when printed in columns because SPSS inserts a blank in front of single-digit numbers.
yy	A two-digit year in the range 00, 01, . . . , 98, 99. The assumed first two digits of the four-digit year this represents are determined by the configuration found at Edit⇨Options⇨Data.
yyyy	A four-digit year in the range 0001, 0002, . . . , 9998, 9999.

You can go back and change the format of a date variable at any time without fear of losing information. For example, you could enter the data under a format that accepted only the year, month, and day, and then change the format to something that contains only the hours and minutes. The format may not display all the information you entered (in fact, in this case, it won't), but when you change the format back to something more inclusive, you will find that all your data is still there.

To enter data, you should choose a format — any format — that contains all the data you have. You can later change to a more limited format that displays only the information you want. But you can't go the other way. If you later choose a format that doesn't leave parts out, you will see the defaults that were inserted by SPSS when you entered the data.

Time Schedule

Sometimes you have data that's gathered at regular intervals, and you need to know the time each data record was gathered. But interval tracking can be more complicated than simple counting. For example, you might need to track information for each new case hourly, for an eight-hour workday, for five workdays each week, for a few months. This repetition pattern is known as the *periodicity* of the data. (Now that's a word you should never try to say out loud in public until you've practiced in private.)

In the simplest terms, the periodicity variable contains the time-stamp value of each case.

Here's the good news. SPSS can not only create your periodicity variables but can also insert the periodic values into the variables for all your cases. To do all this, use the following steps:

1. **Define your variables and enter your data.**

   Do not define any of the periodicity variables — they will be generated later automatically. The other variables and data must be entered first using any of the methods described in Chapters 4 and 5.

2. **In Variable View or Data View, choose Data⇨Define Dates.**

   The window shown in Figure 6-2 appears.

3. **Select the desired periodicity.**

   The interval being defined in this example is once each hour, for an 8-hour day, for a 5-day workweek. The starting week number is 1, the day number is 1, and the hour number is 0. The hour numbers count up to 8, and for each count of 8 hours, the day number increases by 1 until it reaches 5, then the week number increases by 1. Each time a number reaches its maximum, it starts over at the beginning.

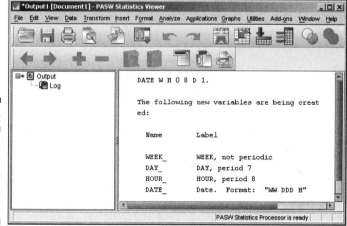

Figure 6-2:
Select the
desired
periodicity
for your
data.

4. Click the OK button.

You're done. The window shown in Figure 6-3 appears, listing the variables that have been defined and added to your previous definitions.
The variable names end with an underscore character (_) to indicate
that they have been generated automatically. You can close this window
if you want — it's only informational.

Figure 6-3:
The list of
variables to
be defined
and popu-
lated with
data.

Figure 6-4 shows the Variable View of the new variables that have been created. The variable named score already existed in the example. The new
variables named WEEK_, DAY_, and HOUR_ are *numeric variables,* used to
hold the numbers of the period. The DATE_ variable is a string data type and
holds a string representation of the value of the other three.

Switch to Data View, and you see the screen shown in Figure 6-5. The first
case was assigned the starting value for each of the new variables, and each
case was assigned the values for the next period in the sequence.

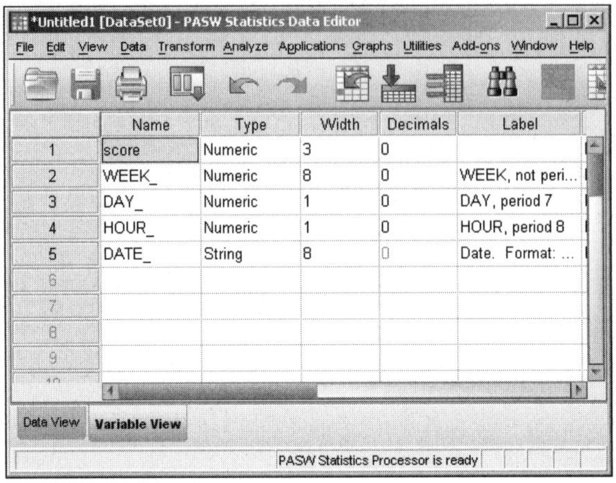

Figure 6-4:
The newly created variables are added to your table.

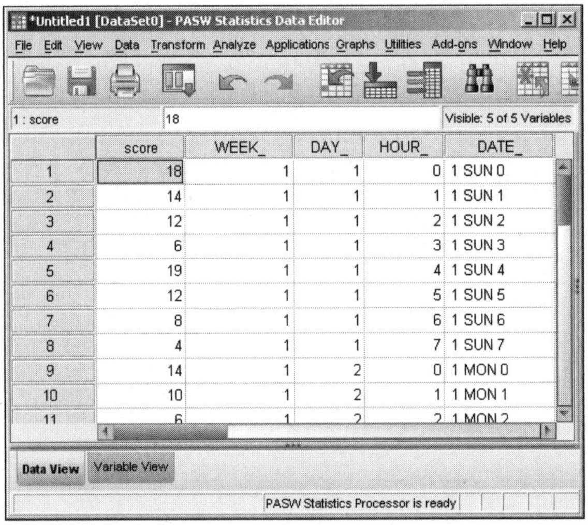

Figure 6-5:
The newly inserted values for the new variables.

Creating a Multiple Response Set

A *multiple response set* is much like a new variable made of other variables you already have. A multiple response set acts like a variable in some ways, but in other ways it doesn't. You define it based on the variables you've already defined, but it doesn't show up in Variable View. It doesn't even show up when you list your data in Data View, but it does show up among the items you can choose from when defining graphs and tables.

The following steps explain how you can define a multiple response set, but not how you can use one — that will come later when you generate a table or a graph.

A multiple response set can contain a number of variables of various types, but it must be based on two or more *dichotomy variables* or two or more *category variables*. For example, suppose you have two dichotomy variables with the value 1 defined as "no" and 2 defined as "yes." You can create a multiple response set consisting of all the cases where the answer to both is "yes," or where the answer to both is "no," or whatever combination you want.

Do the following to create a simple multiple response set:

1. **Create two dichotomy variables that both have 1 for no and 2 for yes as their possible answers, as shown in Figure 6-6.**

   You can do this with more than two variables, but they must all be of compatible types and contain the same set of possible values. (Chapter 4 describes the process of creating variables.)

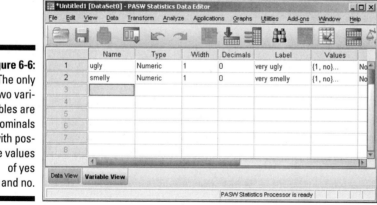

Figure 6-6: The only two variables are nominals with possible values of yes and no.

2. **Choose Data⇨Define Multiple Response Sets.**

   The window shown in Figure 6-7 appears. Your variables appear in the Set Definition area. If you previously defined any multiple datasets, they appear in the list on the right.

3. **In the Set Definition list, select each variable you want to include in your new multiple dataset, and then click the arrow to move the selections to the Variables in Set list.**

   You can move variable names back and forth until you get the list you want. In this example, we need both of them.

4. **In the Variable Coding area, select the Dichotomies option. Specify a Counted Value of 2.**

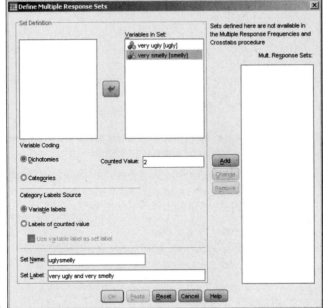

Figure 6-7:
The window
showing all
the informa-
tion about
multiple
response
sets.

With a Counted Value as 2, your new multiple response set will be a count of all the cases in which both variables have the value 2. That is, when you use the variable (for analysis or to draw a graph or whatever), it will only exist where both the dichotomy variables have the value 2. If you get a count of the number of occurrences of the variable, you will have a count of the cases in which the two base variables have a value of 2.

5. Select a Set Name and (optionally) a Set Label.

6. Click Add.

The new multiple response set is created and a dollar sign ($) is placed before the name, as shown in Figure 6-8. The dollar sign in the filename identifies the variable as a multiple response set.

This example used a pair of dichotomy yes/no variables and built a set that counted the cases where both variables were yes. But we could just as well use any group of category variables that all have the same set of answers. For example, if you have the variables Favorite Color, Car Color, and Underwear Color, you could create a multiple response set made up of a count of the instances where all three answers are Red.

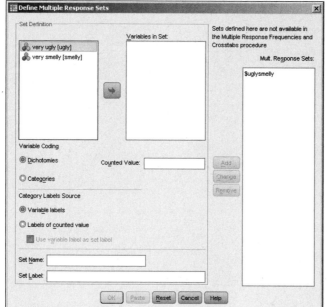

Figure 6-8:
One
response
set has
been
defined.

Copying Data Properties

Suppose you have some data definitions in another SPSS file, and you want to copy one or more of those definitions but you don't want the data. SPSS enables you to choose from several files and to copy only the variable definitions you want into your current table.

If you have a variable *of the same name* defined in your table before you execute the copy, you can choose to change the existing variable definition by loading new information from another file. The copied definition simply overwrites the previous information. Otherwise the copying procedure creates a new variable.

The following steps show you how to copy data properties:

1. **Choose Data⇨Copy Data Properties.**

   The window in Figure 6-9 appears.

2. **Make certain the option for An External SPSS Statistics Data File is selected.**

3. **Click the Browse button, locate the file from which you want to copy variable definitions, and then click Open.**

   The name of the selected file appears next to the Browse button.

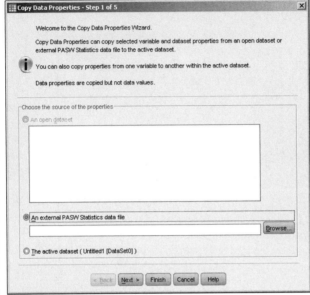

Figure 6-9:
Select the
file you
want to
use as the
source of
variable
definitions.

4. **Click the Next button.**

5. **Select the variables you want.**

   Figure 6-10 displays the variable names that match in the source and destination. In the example, all three are selected, but you can turn the selection of each one on and off: Put the mouse pointer on the one you want to select or deselect, hold down the Ctrl key, and click.

6. **To use the variables you have selected, click Next.**

   If you want to copy the complete definitions of all the variables you've selected and completely overwrite what you have, you can click the Finish button. The Next button, as in this example, allows you to be more specific about which parts of the variable definitions you want to copy.

7. **Choose the properties of the existing variable definitions that you want to copy to the variables you're modifying.**

   In Figure 6-11, everything is selected by default, but you can skip any parts you don't want by deselecting them. These selections apply to all variables you've chosen. If you want to handle each variable separately, you'll have to run through this entire procedure again for each one, selecting different variables each time.

8. **Click Next to be able to select from a list of variable properties.**

If you're satisfied with your choices, you can click the Finish button to complete the process. Clicking Next, as in this example, makes it possible for you to select from a list of available properties to be copied.

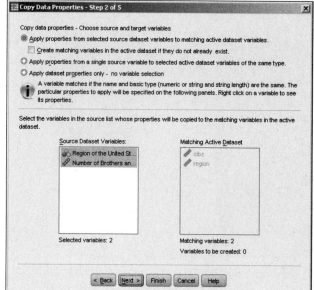

Figure 6-10:
Select the source variable names you want to use for definitions.

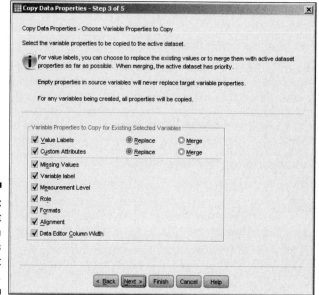

Figure 6-11:
Select which attributes you want to copy.

9. **Choose any properties made available in the dialog box shown in Figure 6-12.**

   Depending on the variable type, different properties are available to be copied. As shown in Figure 6-12, the properties not available appear grayed out. By default, none of them are selected.

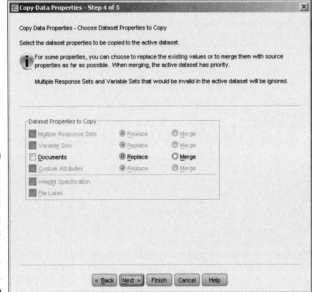

Figure 6-12:
Attributes other than variable definitions can be copied from the source.

10. **Click Next to move to the final dialog box.**

    As shown in Figure 6-13, the screen displays the number of existing variable definitions to be changed, the number of new variables to be created, and the number of other properties that will be copied. You can elect to have the action take place immediately or have the set of instructions saved as a Command Syntax script so you can execute them later. (Part V describes using the Command Syntax language.)

11. **Decide whether to execute the commands now or later.**

    You can click Finish to have the copy procedure execute immediately.

12. **Click Finish.**

Using the basic variable types and the property descriptions you can add, you should be able to concoct any type of variable you need.

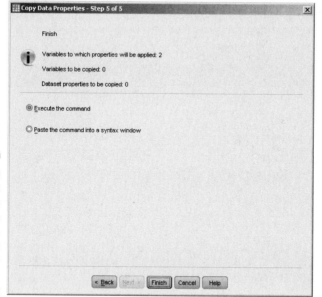

Finish

Variables to which properties will be applied: 2

Variables to be copied: 0

Dataset properties to be copied: 0

◉ Execute the command

○ Paste the command into a syntax window

Figure 6-13:
Choose to
execute the
commands
or save the
commands
for later
execution.

< Back | Next > | Finish | Cancel | Help

Chapter 7

Messing with the Data After It's in There

*A*fter you get your raw data into SPSS, you may find that it contains errors or that it may not be organized the way you'd like. A way to alleviate these problems is by making modifications to your data configuring the values into a form that's easier to work with and to read. This chapter contains some methods you can use to modify your data without loss of information.

Sorting Cases

You can change the order of your cases (rows) so they appear in just about any order you want. You sort them by comparing the values you entered for your variables. The following example uses one of the data files that installs with SPSS. The data will be sorted with all males listed first, and with the youngest males first within that sort order. These two variables — in this example, sex and age — are known as the *primary* and *secondary* sort keys.

You don't need to limit your sorting to two sort keys. You can have a third and fourth key, if necessary, but these keys come into effect only when the keys sorted before them hold identical values. In most cases, two sort keys are plenty to get what you want.

You can sort based on any variables, of any type, by simply selecting the variables as keys. For example:

1. **From the main menu, choose File⇨Open⇨Data and load the** 1991 U.S. General Social Survey **file, which is in the PASW directory.**

   The result is the presentation of a collection of apparently unsorted cases shown in Figure 7-1.

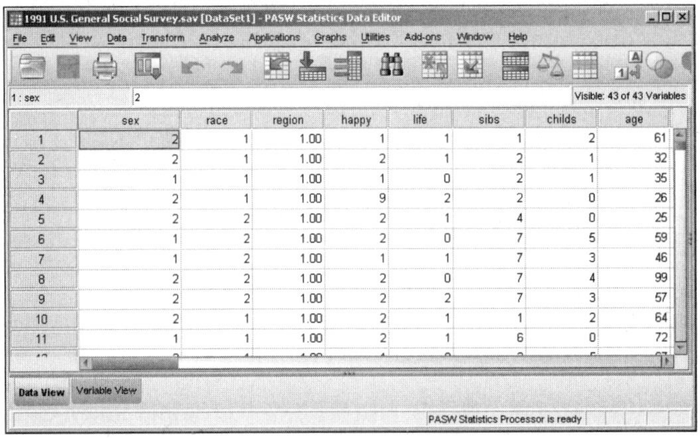

Figure 7-1: The data unsorted, as it is loaded directly from the data file.

2. **From the Data Editor window, choose File⇨New⇨Syntax, and the Syntax Editor window appears.**

3. **In the right panel, to the right of the number 1, enter the four words** SORT CASE SEX AGE. **as shown in Figure 7-2.**

   This is one line of Command Syntax language. Be sure to include the period at the end. Although the command will work without it, SPSS will complain.

4. **From the main menu of the Syntax Editor window, choose Run⇨ To End.**

   The Data View window appears as shown in Figure 7-3. The data has been sorted with the male sex — represented by the number 1 — and the youngest age — which is 18 — at the top of the list of cases. It came up male-first because male is defined as 1, which is a smaller number than the 2 that represents female.

5. **To change the order in which things are sorted, replace the command in the Syntax Editor window with** SORT CASE SEX (D) AGE.

   You can reverse the sort order for any or all variables selected as sort keys. The default is ascending order — smallest to largest — but you can specify descending order by following a variable name with a (D) indicator. The resulting sort, with the youngest female first, is shown in Figure 7-4.

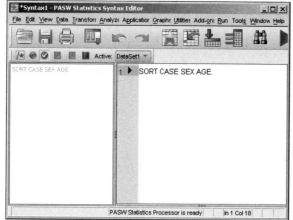

Figure 7-2:
The Syntax Editor window containing a simple syntax.

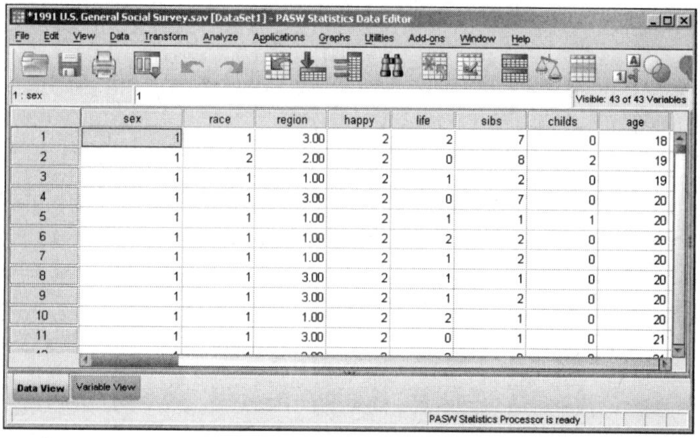

Figure 7-3:
The data sorted with the case of the young-est male first.

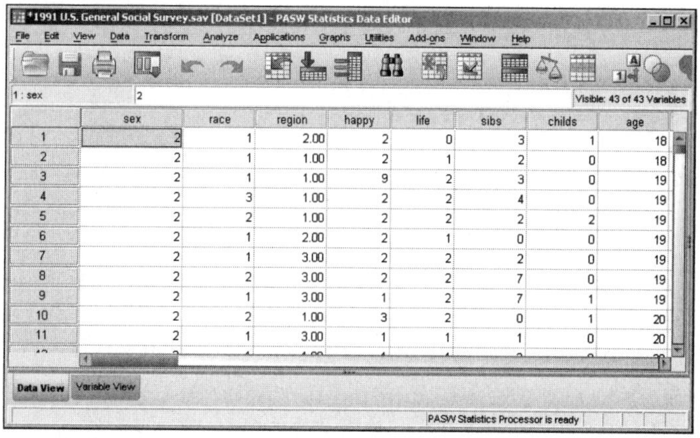

Figure 7-4:
The data sorted with the young-est female first.

Sorting data is strictly for the way you want it to appear in the table. The order in which the data is displayed never affects the analysis.

The order of the sort keys is important. In the preceding example, if AGE had been chosen as the first key and SEX as the second, here's how the sort would have run: All 18-year-olds would have come up first in the list, ordered by female and then male. Following that, the next age would have come up, and it too would have been ordered by sex, and so on.

Counting Case Occurrences

If your data is being used to keep track of multiple similar occurrences — such as people who subscribe to any combination of three different magazines, or eggs produced with something other than a single yolk — you can automatically generate a count of the occurrences for each case. SPSS automates the process of creating a new variable and counting the values for you. You specify what value or values cause a variable to qualify, and SPSS counts the number of qualifying variables from among those you choose. You must have a number of variables that all normally take the same range of values. For example, if you have a number of expenses for each case, you could have SPSS count the number of expenses that exceed a certain threshold.

In the following example, people are listed as subscribers or nonsubscribers to three magazines, which are named simply mag1, mag2, and mag3. The following steps generate a total of the number of subscriptions for each person:

1. **Choose Open⇨File⇨Data and open the** magazines.sav **file.**

   This file can be downloaded as described in the introduction. The screen shown in Figure 7-5 appears.

Figure 7-5:
Each magazine has the value 1 for a subscriber and 0 for a nonsubscriber.

	name	mag1	mag2	mag3	var	var
1	fred	1	0	0		
2	sam	0	1	1		
3	sue	0	0	0		
4	pete	1	1	1		
5						
6						
7						

2. **Choose Transform⇨Count Values Within Cases.**

   The screen shown in Figure 7-6 appears.

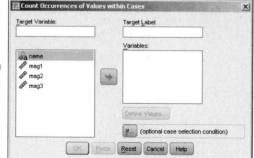

Figure 7-6:
The
initial value-
counting
window.

3. **Select the name of every variable you want to use in the count, and then click the arrow to move them from the panel on the left to the panel on the right labeled Variables. Give your new variable a name.**

   This operation works only with numerics because it must perform numeric matches on the values. If you want, you can come up with both a name *and* a label to be assigned to the variable that this process creates. In this example, the name is `count` and the label is `Count of subscriptions`, as shown in Figure 7-7.

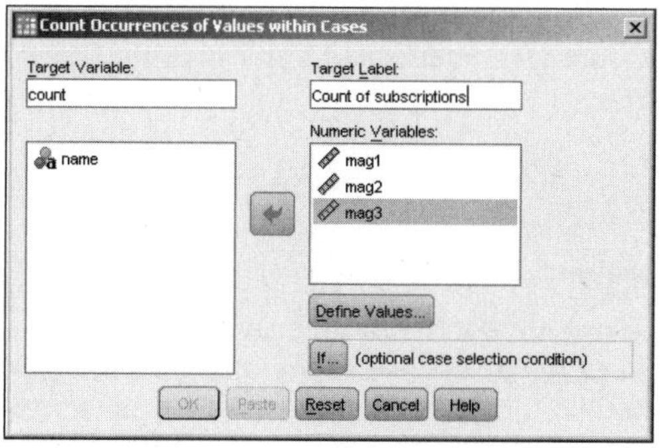

Figure 7-7:
The chosen
variables to
be counted,
and the
name of
the new
variable.

4. **Click the Define Values button.**

   The window shown in Figure 7-8 appears. In this window, I've decided to count, from among the selected variables, those with the numeric value of 1 — which in our example is the value that signifies a subscription.

   As you can see in the figure, the total can also be based on missing values and ranges of values. In the ranges, you can specify both the high and low values, or you can specify one end of the range and have the other end be either the largest or the smallest value in the set. In fact, you can select a number of criteria and SPSS will check each variable against all of them.

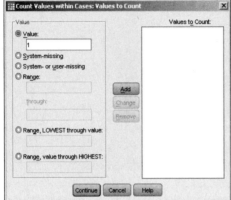

Figure 7-8:
Define the criteria that determine which values are included in the count.

5. **Select a criterion value you want to use, and then click the Add button to move it to the panel on the right labeled Values to Count. Repeat as needed to define all your criteria.**

   The new variable will contain a count of the variables that you named that have a value that matches at least one of the criteria you specified. Each case is counted separately.

6. **Click the Continue button.**

   You return to the Count Occurrences of Values within Cases screen (refer to Figure 7-6).

7. **Click the If button.**

   The window shown in Figure 7-9 appears.

8. **Define your expression.**

   By default, all cases are included, but you can specify criteria here to exclude some cases. To do so, select the Include If Case Satisfies Condition option and, in the text box below, define an expression that specifies the values you want to accept. Then only the values for which the expression is true are considered as candidates for a count greater than 0. You can use any of the variables in the expression. And by using the number pad, the operator buttons, and the function selection, you can construct any expression you want. (For more information on constructing expressions, see Part V.)

9. **Click the Continue button to have SPSS accept your definition. Otherwise (as I did for this simple example), click Cancel and all cases are considered.**

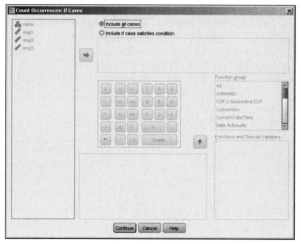

Figure 7-9:
Define
arithmetic
expressions
that deter-
mine which
values are
included in
the count.

10. **Click the OK button and the new field, along with its counts, is generated.**

    The result is the new variable named count, as shown in Figure 7-10.

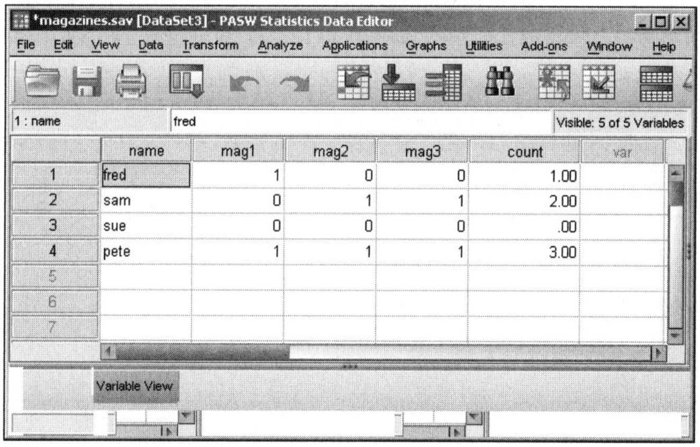

Figure 7-10:
A new
variable
containing
the total
number of
subscrip-
tions per
case.

Recoding Variables

You can have SPSS change specific values to other specific values according to rules you give it. You can change almost any value to anything else. For example, if you have *yes* and *no* represented by 5 and 6, you could recode the values into 1 and 2. You can recode the values in place without creating a new variable, or you can create a new variable and recode values into it. You may want to do this to correct errors or to make the data easier to use.

Recoding into the same variables

When you're recoding values without creating a new variable to receive the new numbers, be sure you store a safety copy of your data before you start. Changes to your data can't be automatically reversed; you could destroy information.

The following example is a list of names of individuals who were invited to an event. If they responded with a yes, the response value was set to 1; if they responded with a no, the value was set to –1. Those with a 0 have not yet responded. As the date of the event approaches, you decide to convert all the –1 responses to 0 to get a count of people not coming. Here's how.

To download the file, go to this book's Web site. You can download this single file or all the files created for this book. Simply place the files in a directory where you can find them through the menus of SPSS. Then follow these steps:

1. **Choose Open⇨File and load the** `rsvp.sav` **file.**

   The window shown in Figure 7-11 appears.

Figure 7-11:
The list of names with the three possible response conditions.

	name	response	var	var	var	var	va
1	fred	-1					
2	sam	0					
3	pete	1					
4	sue	-1					
5	harve	0					
6							
7							

2. **Choose Transform⇨Recode into Same Variables.**

3. **Select the** `response` **variable and click the button with the arrow to move the variable to the panel on the right, labeled Numeric Variables, as shown in Figure 7-12.**

4. **Click the Old and New Values button.**

   The window shown in Figure 7-13 appears.

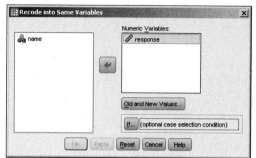

Figure 7-12: A variable name selected to be recoded.

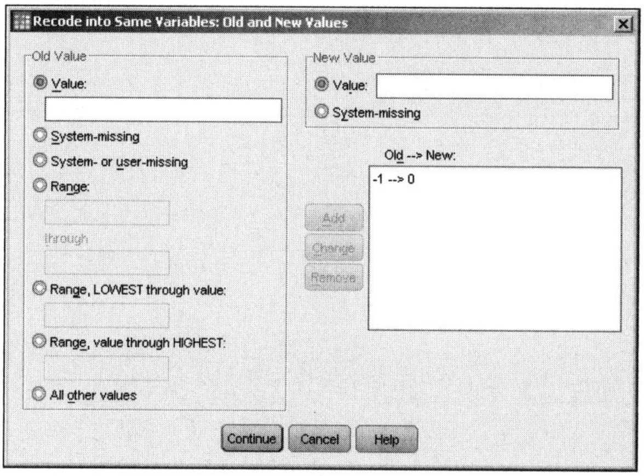

Figure 7-13: Define the recoding of old values into new values.

5. **As shown in the figure, enter an existing value in one of the Old Value choices, and then enter a New Value for it.**

   You can specify a range of old values and map them to a new value. You can also specify that the new value is to be missing and the old value will be mapped to that. You can, if you want, map a number of old values to new values and SPSS will do all the recodings at once. For each mapping of an old value to a new value, use the Add button to make the mapping appear in the window labeled Old–>New.

6. **After you've entered all the mappings (in this example it's just the one), click Continue.**

7. **Optionally, rather than clicking the Old and New Values button as in Step 4, you can click the If button and the window in Figure 7-14 appears, so you can limit the number of cases to which the recoding will apply.**

   You accomplish the limiting by entering an expression that must be true for a case to be included. In our example, we enter no expression, because we want the process to apply to all cases. (For more on expressions, see Part V.)

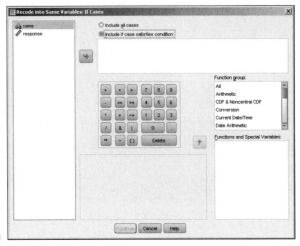

Figure 7-14:
Limit the
cases to
which
recoding
will apply.

8. **Click the OK button.**

All the –1 values are converted to 0, as shown in Figure 7-15. The variable has had its values recoded.

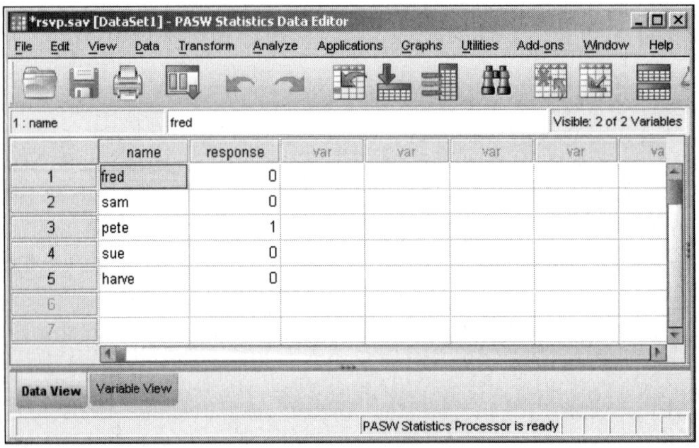

Figure 7-15:
All –1 values
have been
recoded as
0 values.

Recoding into different variables

It could be that you don't want to overwrite the existing values but you'd like to have the recoded data available. The following steps do much the same thing as the preceding example, except the recoded values are stored in a new variable:

1. **With the** `rsvp.sav` **file loaded the same as before (refer to Figure 7-11), choose Transform⇨Recode into Different Variables.**

2. **In the left panel, select the variable holding the values you wish to change. Using the arrow in the center, move the variable name to the panel in the center.**

3. **On the right, in the Output Variable area, enter a name and label for a new variable.**

   For the output variable, you can choose a new variable name (so a new variable is created) or choose an existing variable name and have its values overwritten.

4. **Click the Change button and the output variable is defined, as shown in Figure 7-16.**

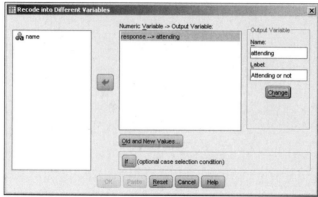

Figure 7-16:
Name the variable to receive the recoded values.

5. **Click the Old and New Values button.**

6. **Define the recoding.**

   Enter an existing value into the Old Value text box and the value you want it to become in the New Value text box. Then click the Add button to add them to the Old–>New list (as shown in Figure 7-17). Be sure to map all values — even the ones that don't change — because you're creating a new variable and it has no preset values.

7. **Click the Continue button.**

8. **Click the OK button.**

   The results appear, as shown in Figure 7-18. Notice that the numbers all have two digits to the right of the decimal point. This may or may not be what you want, but the new variable was created automatically and that is part of the default.

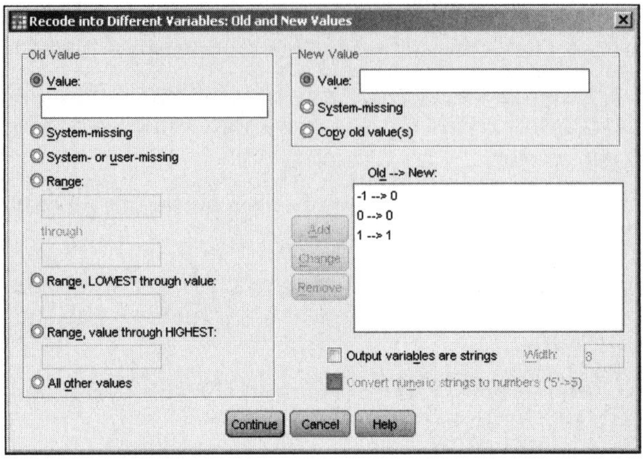

Figure 7-17:
All possible values recoded for a new variable.

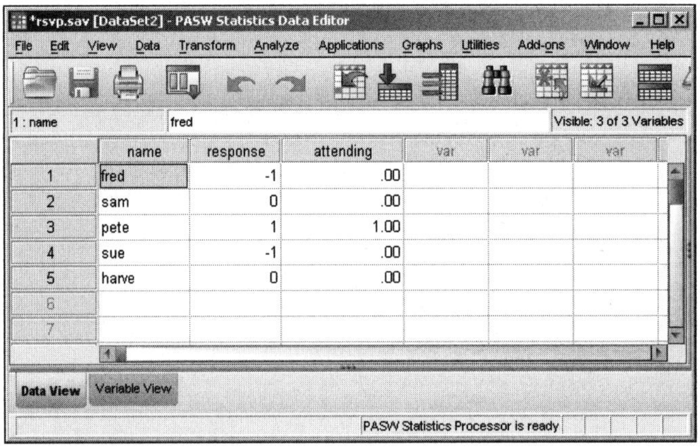

Figure 7-18:
Values recoded into a new variable.

Automatic recoding

Automatic recoding converts values into something you can use in computations. For example, if you have a list of automobile names, automatic recoding converts those names into numbers so you can perform an analysis on the pattern of numbers. Automatic recoding gives you a numeric handle on data that could otherwise elude analysis.

To perform automatic recoding, you select options and set the names in a single dialog box. To see an example of automatic recoding in operation, follow these steps:

1. **Load** `rsvp.sav` **(refer to Figure 7-11).**

2. **Choose Transform⇨Automatic Recode.**

   The Automatic Recode dialog box appears.

3. **In the panel on the left, select the name of the variable you want to recode. Then click the arrow in the middle to move the variable to the panel on the right.**

4. **In the New Name text box, enter the name of the variable to receive the recoded values.**

5. **Click the Add New Name button.**

   The name you entered appears in the panel above the new name, as shown in Figure 7-19.

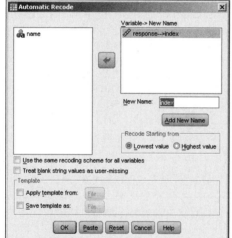

Figure 7-19:
The dialog box for automatic recoding.

6. **Click the OK button and recoding takes place.**

   The result is similar to that shown in Figure 7-20, where the new variable is named `index`.

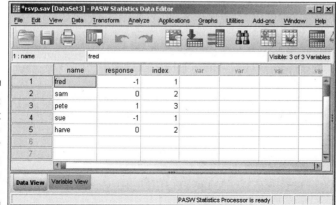

Figure 7-20:
The result of auto-matically recoding name into `index`.

The values in the new variable, index, come about from sorting the values of the original variable and then assigning numbers to them in that order. If the input values are a string of characters instead of the digits of numbers, the strings are sorted alphabetically. (Well, almost: Uppercase letters come before lowercase.)

In the Automatic Recode window (refer to Figure 7-19), you can see the choice for recoding the values with new numbers that start with either the lowest or the highest value. The new numeric values will be the same either way; they're just assigned in the opposite order.

At the bottom of the Automatic Recode window are two choices for the creation of a template file. This is so you can save a file — called a *Template file* — that holds a record of the recoding patterns. That way, if you need to recode more data with the same variable names, the new input values will be compared against the previous encoding and be given appropriate values so that the two data files can be merged and the data will all fit. For example, if you have brand names or part numbers in your data, the recoding will be consistent with the original values because it will be assigned the same *pattern* of recoded values.

Binning

If you're using a scale variable that contains a range of values, you can create groups of those values and organize them into bins. For example, you could use the ages of a number of people and put each one in its own bin — one bin for ages 0 to 20, another bin for 21 to 40, and so on. You can specify the size and content of bins in several ways. The actual binning process is automatic.

The following steps take you through an example of the binning process by dividing salaries into bins:

1. **Choose File⇨Open⇨Data and load the** salaries.sav **file.**

   This file is available for download as described in the introduction. This file contains a list of ID numbers with a salary for each one, as shown in Figure 7-21.

2. **Choose Transform⇨Visual Binning.**

   The dialog box shown in Figure 7-22 appears.

3. **Select Current Salary in the panel on the left, then click the arrow in the center of the window to move the name of the variable to the panel on the right.**

4. **Click the Continue button.**

   A bar graph displaying the range of values of the salaries appears in the center, as shown in Figure 7-23.

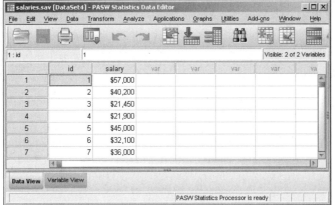

Figure 7-21:
A list of
employee
ID numbers
and the
salaries cor-
responding
to them.

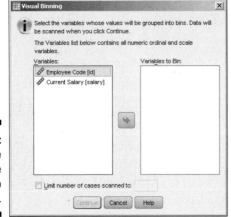

Figure 7-22:
Select the
name of the
variable to
be binned.

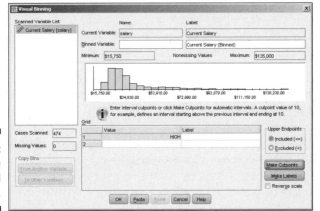

Figure 7-23:
How the
binning will
be done.

5. **Click the Make Cutpoints button.**

   A dialog box appears; here you can specify the size of each bin and the number of bins.

6. **Select the points at which you want to have the data cut into parts to create the bins.**

   In this example, I divided the data into even percentiles of numbers of cases — that is, each bin will contain the same number of cases, as shown in Figure 7-24. Notice that four cutpoints divide the data into five bins, each holding 20 percent of the cases. I could have chosen to divide the data into equal-width intervals — that is, each bin would contain a range of the same magnitude, which would put different numbers of cases in each bin. Also, the cutpoints could have been based on standard deviations, which would create two cutpoints, dividing the data into the three bins — one each of low, medium, and high capacity.

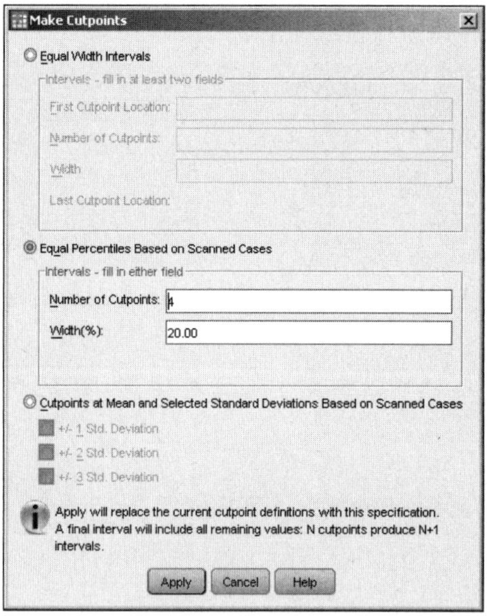

Figure 7-24: Specify how you want the data divided into bins.

7. **Click the Apply button, and the cutpoints appear as vertical lines on the bar graph, as shown in Figure 7-25.**

   You may click the Make Cutpoints button repeatedly and cut the data different ways until you get the cutpoints the way you like. Any new cutpoints you define replace any previous ones.

8. Enter a name for a new variable to contain the binning information.

You enter the name in the Binned Variable text box. The default label for the new variable appears in the text box to the right of the name. You can change this if you want. The bins are created and numbered from 1 to 5, but if you select the Reverse Scale option (in the lower-right corner), the numbering will be from 5 to 1.

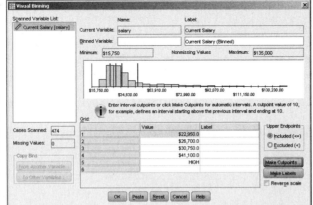

Figure 7-25:
A bar graph of the data with cut-points for binning.

9. Click OK.

The new variable is created and filled with the bin values, as shown in Figure 7-26.

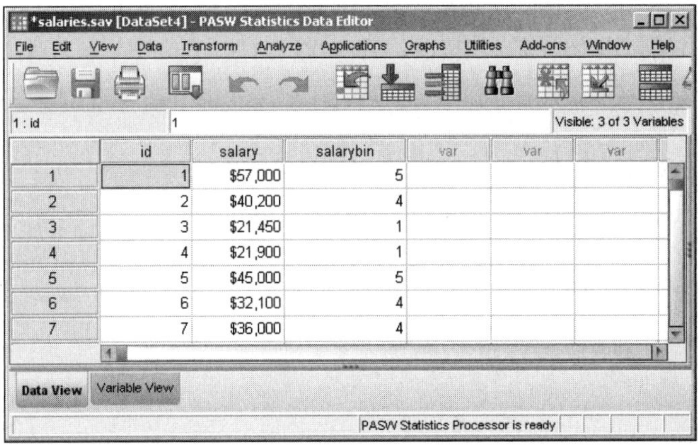

Figure 7-26:
The new variable containing the bin numbers.

The binning is now complete and you can use the new data for further analysis. One thing you can do quickly and easily is display a summary of the contents of your bins. Simply follow these steps:

1. **With the window in Figure 7-26 still on the screen, choose Transform⇨ Optimal Binning.**

2. **Select variable names on the left and click the arrow buttons to move the variables. Move Current Salary to Variables to Bin and move Current Salary (binned) to Optimize Bins with Respect To, as shown in Figure 7-27.**

   The variable in the Optimize Bins with Respect To text box does not have be a variable from a previous binning operation. It can be any variable that contains a collection of values sufficient for being separated into bins.

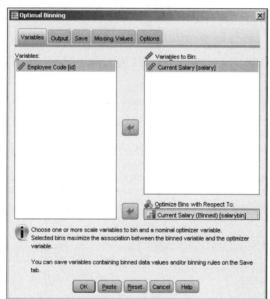

Figure 7-27:
Select the bin variable and the optimizing variable.

3. **Click the OK button.**

   The output is generated, as shown in Figure 7-28.

Current Salary

| Bin | End Point | | Number of Cases by Level of Current Salary (Binned) | | | | | |
	Lower	Upper	1	2	3	4	5	Total
1	a	$23,100	96	0	0	0	0	96
2	$23,100	$26,850	0	95	0	0	0	95
3	$26,850	$30,900	0	0	96	0	0	96
4	$30,900	$41,550	0	0	0	93	0	93
5	$41,550	a	0	0	0	0	94	94
Total			96	95	96	93	94	474

Each bin is computed as Lower <= Current Salary < Upper.
a. Unbounded

Figure 7-28: The output from optimal binning.

Any variable with properly distributed values can be used as the basis of optimal binning. In the chart shown in Figure 7-28, the numbers 1 through 5 across the top are the values of the new binning variable created and stored as part of the data. The numbers 1 through 5 down the left of the graph are the result of the new binning action. The chart lets you see clearly the range of values that make up each bin.

Chapter 8

Getting Data Out of SPSS

. . . .

In This Chapter

▶ Outputting tables and images to the printer

▶ Outputting information to a databases

▶ Outputting tables and images to SPSS Viewer

▶ Outputting to Excel, Word, and other applications

. . . .

SPSS is good at analyzing your data and displaying information that's easy to understand in tables, charts, and graphs — but the time comes when you want to output the results to files suitable for use in other applications. You may want to send output to the printer, or you may have another program that could make use of the output from SPSS. This chapter explains ways that you can output data from SPSS in the particular forms that other programs need.

Printing

The simplest form of output is to print the numeric rows and columns of the raw data as it appears in the Data View tab of the Data Editor window. To do so, choose File⇨Print and a familiar Print dialog box appears, where you can select the print settings you need for your system. The table of data will be printed with lines between the rows and columns, the same as they appear on-screen. The printed form has case numbers to the left and variable names at the top.

If you're not sure what your output will look like, you can choose File⇨ Print Preview and see, on the screen, the same layout that will be sent to the printer. The zoom and page-selection controls at the top of the window allow you to examine the output.

If the table you're printing is too wide to fit on the sheet of paper, SPSS splits the output and places the table on multiple pages. You can hold the printed sheets side by side to get the full width of the table.

If you want to print the variable definitions, you can switch from the Data View tab to the Variable View tab before printout. This output always requires two pages because it includes the full width of the table.

Exporting to a Database

You can export SPSS data directly to a database. Choose File⇨Export to Database and follow the instructions SPSS supplies for your database. SPSS knows how to write to dBase, Excel, FoxPro, Access, and text file databases. If you have a different database system, you should be able to configure SPSS for it by clicking the Add ODBS Data Source button. You should be able to get the information you need to do this from the documentation of your database. In similar fashion, you can read data from a database by choosing File⇨Open Database.

To export the data, simply follow the on-screen instructions for selecting the variables to be written and for choosing whether to append new data or to overwrite existing data.

Using SPSS Viewer

Whenever you run an analysis, produce a graph, or do anything that generates output (even loading a file), the SPSS Viewer window pops up automatically to display what you've created. This display is the most fundamental form of output from SPSS and is the first step in producing other forms of output.

Chapters 9 through 14 provide details about generating tables, graphs, and descriptive text in the SPSS Viewer window. These chapters describe how to output Viewer data to files in different formats.

You can output data from SPSS Viewer in several file formats appropriate for use by other applications. Some output formats are graphics only, some are text only, and others are a mixture of text and graphics. Some form of graphic output is usually necessary because of the graphs and charts constructed by SPSS.

In every case, you begin by choosing File⇨Export from the menu of the SPSS Viewer, which displays the Export Output dialog box (shown in Figure 8-1). In the Export drop-down list, you can choose which items in the View window to export — the entire document, the text of the document without graphics, or the graphics without text.

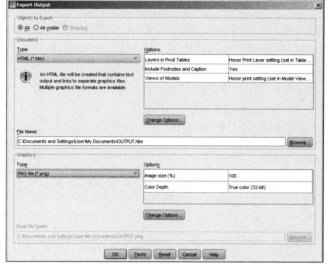

Figure 8-1:
The main
control
window for
generat-
ing output
from SPSS
Viewer.

At the very top of the dialog box, you can select which pieces of information in SPSS Viewer you want to include as part of the output:

- ✔ **All:** This option outputs all the information that SPSS Viewer contains, whether or not the information is currently visible.

- ✔ **All Visible:** This option includes only those objects being displayed by SPSS Viewer.

- ✔ **Selected:** This option allows you to select which objects to output.

The set of selections made available to you in Export Output is determined by the types of objects being displayed by SPSS Viewer, which (if any) are selected, and the choice in the Export drop-down list. The only combinations of options available are those that produce output.

Figure 8-2 shows a SPSS Viewer window displaying both text and graphics. On the left is a list of names of objects. If the name of an object is visible in the list, the object itself is visible in the Viewer window. You can make objects appear and disappear by clicking the plus and minus signs. If the name of an object is highlighted in the list, the object is marked as selected in the Viewer window; a selected object appears surrounded by boxes. (In the figure, the log, title, and notes at the top are not selected, but the other objects are.) When producing output, you can select to export only visible objects, only selected objects, or all objects.

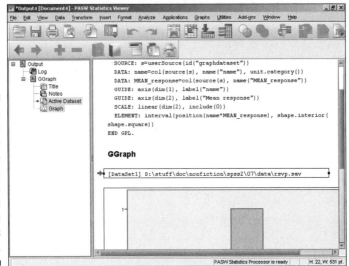

Figure 8-2:
A SPSS
Viewer
window
displaying
text and
graphics
with some
objects
selected.

You can output the following types of files:

- Plain text
- Unicode (UTF8 or UTF16)
- HTML Web page
- Excel file
- Rich text format (RTF), readable by Word
- PowerPoint display file
- Portable document format (PDF)

Some formats (for example, the text-file format) require that graphics be output in separate files; you can also elect to output *only* graphics files. Graphics can be output in the following formats:

- Standard jpeg (JPG)
- Portable network graphics (PNG)
- Postscript (EPS)
- Tagged image file format (TIFF)
- Windows bitmap (BMP)
- Enhanced metafile (EMF)

Creating an HTML Web page file

If you decide to format your output file as a Web page, the output text will be formatted as HTML. Any pivot tables selected for output will be formatted as HTML tables, and any images to be output will be written to separate files in the image format of your choice (a description of the image-file options appears at the end of this chapter).

You can make a number of decisions about the details of the HTML file, as shown in Figure 8-3; they appear when you click the Change Options button for the Document in the Export Output dialog box.

Figure 8-3:
The options
for creating
an HTML
file.

The first options to set are the layers in pivot tables. Some pivot tables have more than two dimensions, and the other dimensions are presented as multiple displayable layers. By setting this option, you can include or exclude layers in the HTML file — if you have a multilayered table, you'll probably have to experiment with this setting to get desirable results.

A pivot table can have multiple headings and footnotes. You can choose to have the footnotes and headings included or excluded.

The on-screen view is not the only one available. You have the option of including all views in the output or showing only the view that is currently visible.

Figure 8-4 shows part of the output page as it appears in a Web browser — using the default settings for everything, including the JPEG image. Notice that the commands that generated the graphics were included and formatted in an HTML table. You may decide to leave that information out. You could, if you want, leave the table out and include only the graphic and its annotation. Also, if you were going to publish this as a Web page, you would probably want to edit the heading so it's something other than the name of a working directory on your local machine.

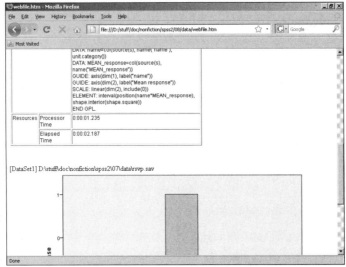

In this example, the output filename is webfile, so the main file is webfile. htm and the image file is webfile1.jpg. The jpg suffix indicates that a JPEG image file was the chosen option. The digit in the image filename is necessary because there could be more than one; each needs a unique name.

Creating a text file

If you want to output a simple text file, you still have a number of options to choose from, as shown in Figure 8-5. The first two options are whether to use spaces or tabs to position characters on the page. This choice can be important because alignment is crucial to some data layouts, and programs that read the text files might have different tab settings and change the appearance of the output when it's displayed.

The options for creating Unicode file are the same as those for creating a plain-text file. The Unicode output is in the standard encoding format of your choice — either UTF8 or UTF16. You would only want to output text in one of these formats if you have a program that needs one of those formats for its input.

Tables output as text use certain characters to define the cells in which data items are shown. You can select any characters you want to act as separators and draw the borders, or you can accept the default of the minus sign and vertical bar, as shown in the figure. (The vertical bar is a standard keyboard character, usually on the same key as the backward slash. It sometimes looks like a vertical line broken in the middle.) If you're outputting tables, you can choose a maximum cell size or just use the default Autofit option and let SPSS decide the number of characters that will fit in each column.

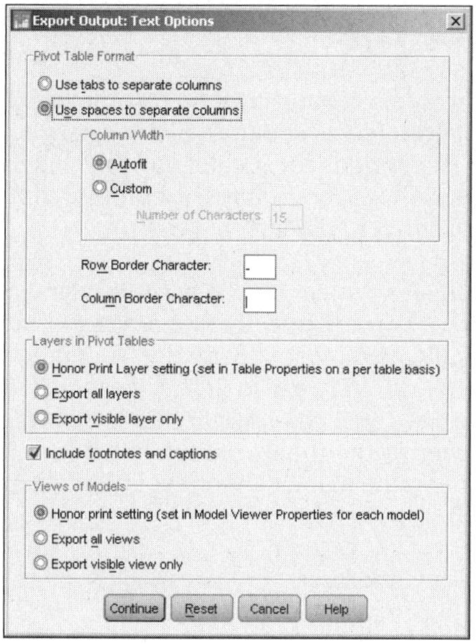

Figure 8-5:
The options
for creating
a text file.

The output shown in Figure 8-6 is a simple listing in a DOS command-line window of a text file generated by SPSS. It is the same data as in the previous example, which was formatted into HTML. Also, like HTML, the graphic is output in a separate file. The text file includes the full path name of the produced graphic file. You have the same set of options for producing graphic files as you have for Web page files.

Figure 8-6:
SPSS output
as a text file.

In this example, the output filename is `textfile`, so the main file was named `textfile.txt` and the graphic file was named `textfile1.png`. The `png` suffix indicates that a PNG graphic file was the chosen option. As with all such choices, the digit in the graphic filename is necessary because there could be more than one, and each needs a unique name.

Creating an Excel file

Creating an Excel file is easier than creating either a text file or HTML file because the images are not generated as separate files — graphic images are included in the worksheet. (The options for creating an Excel file are shown in Figure 8-7.) You get to choose how pivot tables, footnotes, captions, and models are handled, as described for HTML files.

By default, a new workbook file is created. If a file with the same name already exists, it is overwritten. You can specify that a new worksheet be created within the workbook file; you must specify the worksheet name. If the worksheet name you choose already exists in the workbook, the file that has it is overwritten. Alternatively, you can specify that the output be used to modify an existing worksheet within the workbook file. If you decide to specify the name of a worksheet, the name cannot exceed 32 characters and should not include any special characters.

Also, if you choose to modify a worksheet, you can specify where, in the existing worksheet, the new information is to be placed.

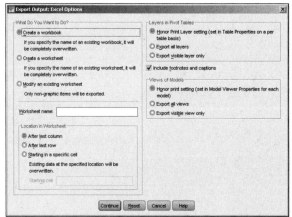

Figure 8-7:
The choices you have when producing an Excel file.

When you want to produce output, click the OK button in the Export Output dialog box, and a file is generated. Then you can load the file directly into Excel, as shown in Figure 8-8.

In this example, the output filename is `excel`, so the output file was named `excel.xls`.

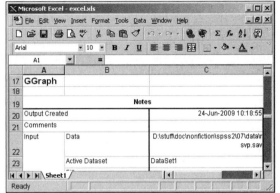

Figure 8-8:
SPSS output
as an Excel
file.

Creating a Word document file

If you choose to output a Word document file, you have no graphic options
to set because both text and graphics are included in one output file. The
options you can choose from are shown in Figure 8-9: whether to include
footnotes and captions, how models are to be handled, and whether to
include all layers of any tables that may be in the output.

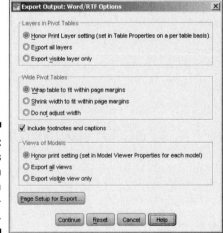

Figure 8-9:
The choices
you have in
producing a
Word docu-
ment.

The Page Setup for Export button opens a dialog box that allows you to lay
out the page size and margins of the output. It makes it possible to specify
wrapping and shrinking to make things fit.

When you want to produce output, click the Continue button in the Export Output dialog box, and the file is generated. Then you can load the output file directly into Word, as shown in Figure 8-10.

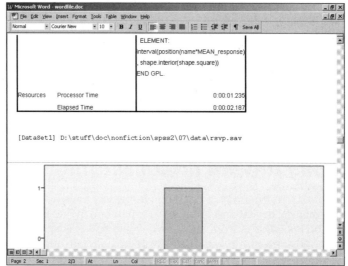

Figure 8-10:
SPSS output
as a Word/
RTF file.

The output file is in RTF (rich text format), a file type that can be loaded and used by most word processors, including OpenOffice, StarOffice, and WordPerfect.

In this example, the output filename is `wordfile`, so the output file was named `wordfile.doc`.

Creating a PowerPoint slide document

A PowerPoint file includes only tables, graphs, and models, so you can produce a series of display slides that contain all your graphics. The basic options are shown in Figure 8-11.

The first options to set are the layers in pivot tables. Some pivot tables have more than two dimensions, and the other dimensions are presented as multiple display layers. By setting this option, you can include or exclude layers in the PowerPoint slides. If you have a multilayered table, you'll probably have to experiment with the setting and see what you get.

A pivot table can have multiple headings and footnotes, which you can choose to include or exclude. You can also choose to use the outline headings as slide titles in your output.

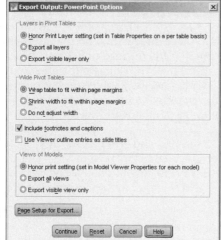

Figure 8-11:
The options
for pro-
ducing
PowerPoint
slides.

Some results are presented as graphic models, but the view that appears on-screen is not the only one you can use. You have the option of including all views in the output, or showing only the view that is currently visible.

Your output will include only charts, graphs, and pivot tables; the rest of your data is ignored and doesn't appear anywhere in the set of produced slides. Figure 8-12 displays the slide produced from the same SPSS Viewer data that was used in the previous example. If you need some text slides before or after your graphics, you have to add those yourself.

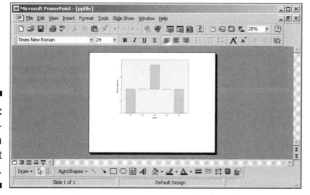

Figure 8-12:
SPSS out-
put as a
PowerPoint
slide.

In this example, the output filename is ppfile, so the output file was named ppfile.ppt.

Creating a PDF document

It is becoming more common to place information on the Internet in a PDF format instead of an HTML format. Both are read-only files, but a PDF gives the creator of the file more control over the document's appearance when it's displayed in a viewer. An HTML page is relatively free-form compared to a PDF file. With a PDF file, you can put your information on the Internet and have it seen the same way by every person who views it.

A PDF file contains both formatted text and graphics, so any PDF you output will look very much like the original data displayed in SPSS Viewer. PDF handles graphics in a standard way, so you don't have the typical graphic options to set. Note, however, that you do have some other options, as shown in Figure 8-13.

Figure 8-13: The options for producing a PDF file.

You can elect to include bookmarks in the produced file. These bookmarks are important for larger files. They are used by the viewer to simplify the process of navigating through the file.

Embedding fonts ensures that the document will look the same on every computer. If the fonts are not embedded, the chosen font may not be available for display or print, in which case the substitute font could make the resulting display look quite different.

You can set the layers in pivot tables. Some pivot tables have more than two dimensions, and the other dimensions are presented as multiple displayable layers. By setting this option, you can include or exclude layers in the PDF file. If you have a multilayered table, experiment with this setting until you get the results you want.

Using the default settings, SPSS produced the PDF file shown in the Adobe Acrobat Viewer in Figure 8-14.

Figure 8-14:
SPSS output displayed by a PDF viewer.

Creating a Graphics File

Depending on the type of output data file you generate, you may need to select the file type and configuration settings for separate image files. When you produce such image files, you could get several of them — one for each image displayed in the SPSS Output Viewer. And image files have options.

If you don't have an immediate handle on the options you can use to generate your selected type of graphics file, experiment. Start with the defaults and make changes only if you need to. It doesn't cost anything to try different combinations of options and decide on the settings you like.

For all image file types, you can specify the size in terms of a percentage of the original. The default is 100 percent, which means there is no change in the size of the image. The other options available (compression, number of colors, and so on) vary, depending on the file type.

Figure 8-15 is the dialog box used to set the options for a bitmap (.bmp) file. The size can be expanded up to 200 percent of the size displayed in the SPSS Output Viewer. You can also choose to use compression to reduce the size of the file — the compression used will not reduce the quality of the image.

Figure 8-15:
Options for
configuring
.bmp files.

Figure 8-16 is the dialog box used to set the options for an enhanced metafile (.emf). The only option is to adjust the size of the image.

Figure 8-16:
Options for
configuring
.emf files.

Figure 8-17 is the dialog box used to set the options for an encapsulated postscript (.eps) file. You can set the size of the image in one of two ways — you can set it to be a percentage of the current size or you can set its width as a number of points. (There are 72 points to an inch.) You can optionally choose to produce a TIFF along with the postscript image, in case you are unable to display the postscript image. If the fonts are all available on the output device, you can simply include the font information. If the fonts are not available, a substitute will be chosen. Alternatively you can choose to present the fonts as a collection of graphics (curves).

Figure 8-17:
Options for
configuring
.eps files.

Figure 8-18 is the dialog box used to set the options for a JPEG (.jpg) file. You can set the size, or you can choose to remove color from the image.

Figure 8-18:
Options for
configuring
.jpg files.

Figure 8-19 is the dialog box used to set the options for a PNG (.png) file. You can set the size as a percentage of the original. The color depth determines the maximum number of colors that can be used in the display. If the output is composed of fewer colors than appear in the original, the output is dithered to differentiate the graphics.

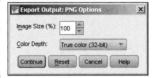

Figure 8-19:
Options for
configuring
.png files.

Figure 8-20 is the dialog box used to set the options for a tagged image file (.tif). The only option is to set the size of the image.

Figure 8-20:
Options for
configuring
.tif files.

Part III
Graphing Data

The 5th Wave   By Rich Tennant

"Look—what if we just increase the size of the charts?"

In this part . . .

Data displayed as a graph often makes more sense than data displayed as columns or tables of numbers. In SPSS you can find lots of different kinds of graphs, some more suitable than others for displaying your particular set of data.

SPSS makes it easy to display data in different graphical formats; you can choose the one you like. You do the clicking and SPSS does the formatting.

Chapter 9

Fundamentals of Graphing

· ·

· ·

*O*ver the years, the SPSS software has improved its methods for generating graphic displays of data. You can take the easy way and be guided through every step, or you can take a faster way and simply enter the values needed to build the graph you want. The older methods of producing graphs as output are still available and on the menu, so if you *like* to suffer while you work, you can use the procedures developed in previous years. In any case, you never have to worry about the size of text and graphics, and you don't have to think about the placement of the graph on the page — SPSS does all the grunt work for you.

SPSS can display your data in a bar chart, a line graph, an area graph, a pie chart, a scatterplot, a histogram, a collection of high-low indicators, a box plot, or a dual-axis graph. Adding to the flexibility, each of these basic forms can have multiple appearances. For example, a bar chart can have a two- or three-dimensional appearance, represent data in different colors, or contain simple lines or I-beams for bars. The choice of layouts is almost endless.

In the world of SPSS, the terms *chart* and *graph* mean the same thing and are used interchangeably.

The Graphs menu in the SPSS Data Editor window has three options: Chart Builder, Graphboard Template Chooser, and Legacy Dialogs. These options are different ways of doing the same job. If you prefer to build graphs in SPSS the original way, of course, you can choose Legacy Dialogs. A better way of building graphs was devised a few years later — the Graphboard Template Chooser — and when it hit the scene, the original way of building graphs became known as Legacy. A few years later, an even better procedure for building charts was devised and added to the menu — Chart Builder. All three building methods are in place primarily for people who are in the habit of using the older procedures, but if you build a lot of graphs, you may find advantages and uses for all of them. You can get the same graphs from all three; only the process is different.

Building Graphs the Easy Way

SPSS contains Chart Builder, which uses a graphic display to guide you through the steps of constructing your display. It checks what you're doing as you proceed and won't allow you to use things that won't work. If the OK button is available for clicking after you've defined what you want as a result, that means that everything is ready and a chart will be produced.

Gallery tab

The following example steps you through the process of creating a bar chart, but you can use the same fundamental procedure to build a chart of any design. You can follow this tutorial once to see how it all works. Later on you can use your own data and choices.

You can't hurt your data by generating a graphic display. Even if you thoroughly mess up the graph, you can always redo it without fear. This is one place where mistakes don't cost anything. And nobody's watching.

The following steps build a bar chart:

1. **Choose File⇨Open⇨Data and load the** 1991 U.S. General Social Survey.sav **file, which is in the PASW directory.**

2. **Choose Graphs⇨Chart Builder.**

   The Chart Builder dialog box appears, as shown in Figure 9-1. If a graph was generated previously, the display will be different, and you will need to click the Reset button to clear the Chart Builder display.

3. **Make certain the Gallery tab is selected.**

4. **In the Choose From list, select Bar as the graph type.**

   The fundamental types of bar charts appear in the gallery to the right of the list.

5. **Define the general shape of the bar graph to be drawn.**

   You can do so in two ways. The simplest is to choose one member from the set of diagrams of bar graphs appearing immediately to the right of the list. For this exercise, select the diagram in the upper-left corner and drag it to the large chart preview panel at the top. Alternatively, you can click the Basic Elements tab (instead of the Gallery tab) and drag one image from each of the two displayed panels to the panel on top, which constructs the same diagram as the bar graph. Figure 9-2 shows the appearance of the window after the dragging is complete. The result is the same no matter which procedure you follow.

You can always back up and start over: At any time during the design of a graph, click the Reset button. Anything you dragged to the display panel is deleted, and you can start from scratch.

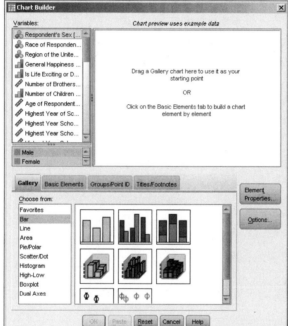

Figure 9-1: The initial Chart Builder window with Bar chosen.

6. **Click Close to close the Element Properties window (see Figure 9-3).**

   This window should have popped up when you dragged the graphic layout to the panel. This dialog box is not needed for this example, so you can close it. If it didn't appear but you'd like to see it, you can click the Element Properties button at any time.

7. **From the list on the left, select the variable with the label and name Highest Year of School Completed (Educ) and drag it to the Y-Axis label in the diagram.**

8. **In similar fashion, select the variable with the label and name Region of the United States (region) and drag it to the X-Axis label in the diagram.**

   The screen now looks like the one shown in Figure 9-4.

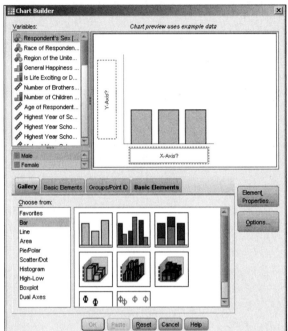

Figure 9-2:
The appear-
ance of the
new bar
chart is
defined.

Figure 9-3:
Use the
Element
Properties
window to
modify chart
elements.

The graphics display inside the Chart Builder window *never* represents your actual data, even after you insert variable names. This window simply displays a diagram that demonstrates the composition and appearance of the graph that will be produced.

Figure 9-4:
The diagram after assigning the X- and Y-axes.

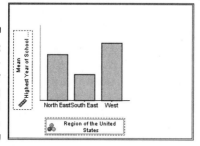

9. Click the OK button to produce the graph.

A SPSS Viewer window appears, containing the graph shown in Figure 9-5. This graph is based on the actual data; it shows that the average number of years of education varied little from one part of the country to the next in this survey.

Figure 9-5:
A bar chart produced from a data file and displayed by SPSS Viewer.

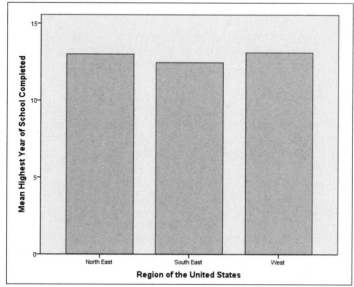

These steps demonstrate the simplest way possible of generating a chart. Most of the options available to you were left out of the example so it would demonstrate the simplicity of the basic process. The following sections describe the options.

Basic Elements tab

The example in the preceding section used the Gallery tab to select the type and appearance of the chart. Alternatively, you can click the Basic Elements tab in the Chart Builder dialog box and select one part of the chart from each of the two panels shown in Figure 9-6.

Figure 9-6:
Choose the axes and elements to construct the graph you want.

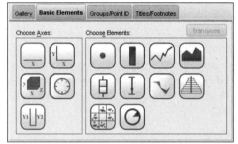

It's sort of like the menu in a Chinese restaurant — you choose one from column A and another from column B. You drag one image from each panel into the panel at the top, and they combine to construct a diagram of the graph you want.

The result is the same as you get from using the Gallery tab. The only difference is that you use the Basic Elements tab to build the graph from its components. Whether you use this technique or the Gallery depends on your conception of the graph you want to produce.

Groups/Point ID tab

Once you have selected the type and appearance of your chart through either the gallery or the Basic Elements tab, you can click the Groups/Point ID tab in the Chart Builder dialog box, which provides you with a group of options you can use to add another dimension to your graph.

In the example in Figure 9-7, I selected the Rows Panel Variable option, which generates a multifaceted graph. The new dimension adds a separate graph for the number of children in the family. A separate set of bars is drawn for those with no children, another set for those with one child, another for those with two children, and so on.

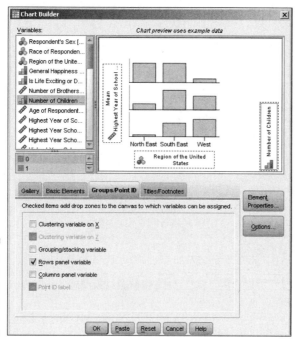

Figure 9-7:
You can add
dimensions
to your
graph.

The Columns Panel Variable option enables you to add a variable along the other axis, thus adding another dimension. Adding variables and new dimensions this way is known as *paneling,* or *faceting.*

Clustering (gathering data into groups) can also be done along the X- or Y-axis if the variables are the type that will cluster (or bin) properly.

Titles and footnotes tab

Figure 9-8 shows the window you get when you click the Titles/Footnotes tab in the Chart Builder dialog box. Each option in the bottom panel places text at a different location on the graph. When you select an option, the Element Properties window appears so you can enter the text for the specified location.

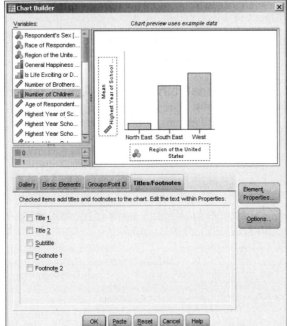

Figure 9-8:
Select the
chart's text
and its
location.

Element Properties dialog box

You can use the Element Properties dialog box at any time during the design of a chart to set the properties of the individual elements in the chart. One mode of the dialog box is shown in one mode in Figure 9-3, and another, in Figure 9-9, changes every time you choose a different member from the list at its top.

The dialog box often pops up on its own when you add an item to the graph's definition. You can make it appear any time you want by clicking the Element Properties button in the Chart Builder dialog box.

Okay, the upcoming list of options is long, but four facts make them simple to use:

- ✔ **All options have reasonable defaults.** You don't have to change any of them unless you want to.

- ✔ **You can always back up and change whatever settings you made.** Nothing is permanent, so you can make changes until you've finished or run out of time and decide, "That's good enough."

- ✔ **Not all options appear at once.** Only a few show up at a time. In fact, you'll probably never see some of the possible options.

✓ **All options become obvious when you see what they do.** You don't have to memorize any of them, but you'll find they are easy to remember.

The following is a simple explanation of all the possible options that can appear in an Element Properties dialog box:

✓ **Edit Properties Of:** This list, which appears at the top of the window, is used for selecting which element in the chart you want to edit. Each element has a type, and the type of the element you select determines the other options available in the window. The selected element is also highlighted in the diagram of the graph in some way.

✓ **X:** When an element is selected and the X button to the right of the list becomes enabled, clicking the button removes the element from the list and from the graph.

✓ **Arrow:** For charts with dual Y-axis variables, the arrow to the right in the list indicates which of the variables will be drawn on top of the other. You can click the arrows to change the drawing order.

✓ **Statistics:** For certain elements, you can specify the statistics (that is, the type of value) to be displayed in the graph. For example, you can select Count and use simple numeric values. You can also select Sum, Median, Variance, Percentile, or any of up to 32 statistic types. Not all types of charts have that many options; the options that are available also depend on the types of variables you're using. For certain statistics options — such as Number in Range and Percentage Less Than — the Set Parameters button is activated; you have to click it to set the parameters controlling your choice.

✔ **Axis Label:** You can change the text used to describe a variable. By default, the variable's label is used.

✔ **Automatic:** If selected, the range of the selected axis is determined automatically to include all the values of the variable being displayed along that axis. This is the default.

✔ **Minimum/Maximum:** You can replace the Automatic default values and choose the extreme values that determine the starting and ending points of an axis.

✔ **Origin:** Specifies a point from which chart information is graphed. This option has different effects for different types of charts. (For example, choosing an origin value for a bar chart can cause bars to extend both up and down from a center line.)

✔ **Major Increment:** The spacing that determines the placing of tick marks, along with numeric or textual labels, on an axis. The value of this option determines the interval of spacing when you also specify minimum and maximum values.

✔ **Scale Type:** You have four different types of scale you can use along an axis:

   • **Linear:** A simple, ruler-like scale. This is the default.

   • **Logarithmic (standard):** Transforms the values into logarithmic values for display. You can also select a base for the logarithms.

   • **Logarithmic (safe):** Same as standard logarithms, except the formulas that calculate values can handle 0 and negative numbers.

   • **Power:** Raises the values to an exponential power. You can select an exponent other than the default value of 0.5 (which is the square root).

✔ **Sort By:** You can select which characteristic of a variable will be used as the sort key. It can be one of the following three:

   • **Label:** Nominal variables are sorted by the names assigned to the values; you can choose whether to sort in ascending or descending order.

   • **Value:** Uses the numeric values for sorting. You can choose whether to sort in ascending or descending order.

   • **Custom:** Uses the order specified in the Order List.

✔ **Order List:** The list of possible values is flanked by up and down arrows. You can change the sorting order by selecting a value and clicking an arrow to move the selection up or down. To remove a value from the produced chart, select its name in the list and click the X button; the value moves to the Excluded list. When you change the Order List, then Sort By switches automatically to Custom.

✔ **Excluded:** Any value you want to exclude from the Order List appears in this list. To move a value back to the Order List (which also causes the value to reappear on the chart), select its name and click the arrow to the right of the list.

If a value (or a margin annotation representing a value) is unexpectedly missing from a graph based your selections, look in this Excluded list. You may have excluded too much.

✔ **Collapse:** If you have a number of values that seldom occur, you can select this option to have them gathered into an "Other" category. You specify the percentage of the total number of occurrences to make it an "Other" value.

✔ **Error Bars:** For Mean, Median, Count, and Percentage, confidence intervals are displayed. For Mean, you must choose whether the error bars will represent the confidence interval, a multiple of the standard error, or a multiple of the standard deviation.

✔ **Bar style:** You can choose one of three possible appearances of the bars on a bar graph.

✔ **Categories:** You can choose the order in which the values appear when they're placed along an axis. You can select ascending or descending order. If the variable is nominal, you can select the individual order and even specify values to be left out.

✔ **Small/Empty Categories:** You can choose to include or exclude missing value information.

✔ **Display Normal Curve:** For a histogram, you can choose to have a normal curve superimposed over the chart. The curve will use the same mean and standard deviation values as the histogram.

✔ **Stack Identical Values:** For a chart that will appear as a *dot plot* (a pattern of plotted points), you can choose whether points at the same location should appear next to one another or one on top of the other (that is, with one point blotting out the one below it).

✔ **Display Vertical Drop Lines between Points:** For a chart that will appear as a *dot plot* (a pattern of plotted points), any points with the same X-axis values show a vertical line joining them.

✔ **Plot Shape:** For a dot plot, you can choose

- **Asymmetric:** Stacks the points on the X-axis. This is the default.

- **Symmetric:** Stacks the points centered around a line drawn horizontally across the center of the screen.

- **Flat:** The same as Symmetric, except no line is drawn.

✔ **Interpolation:** For line and area charts, the algorithm used to calculate how the line should be drawn between points:

  • **Straight:** Draws a line directly from one point to the next.

  • **Step:** Draws a horizontal line through each point; the ends of the horizontal lines are connected with vertical lines.

  • **Jump:** Draws a horizontal line through each point, but the ends of the lines are not connected.

  • **Location:** For Step and Jump interpolation; using this option adds an indicator at the actual point.

  • **Interpolation through Missing Values:** For Straight, Step, or Jump, this option draws lines through missing values. Otherwise the line shows a gap.

✔ **Anchor Bin:** The starting value of the first bin. This option is available for histograms.

✔ **Bin Sizes:** Sets the sizes of the bins when you're producing a histogram.

✔ **Angle:** Rotates a pie chart by selecting the clock position at which the first value starts. You can also specify whether the values should be included clockwise or counterclockwise.

✔ **Display Axis:** For a pie chart, you can choose to display the axis points on the outer rim.

Options

Clicking the Options button in the Chart Builder dialog box opens the Options dialog box, shown in Figure 9-10.

When you define the characteristics of a variable, you can specify that certain values be considered missing values. The options in the Break Variables area let you decide whether you would like those included or excluded from your chart. You can also specify how you would like summary statistics handled. (Missing values are discussed in Chapter 4, and the different types of summary data are described in Chapter 7.)

Templates are files that contain all or part of a chart definition. You can insert one or more names of template files into the list in this window, and SPSS will apply those template definitions as the default starting points for all charts you build. You create a template file from a finished chart displayed in SPSS Viewer. You find out more on making templates later in this chapter.

Templates come in handy only when you want to build lots of similar charts. You can use the Chart Size option to make the generated charts smaller or larger, as needed.

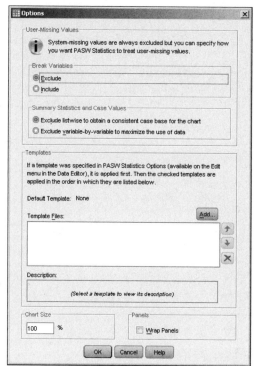

Figure 9-10:
Options you
can apply to
a chart.

The Wrap Panels option determines how the panels are displayed when you have a number of them in a chart. SPSS is using the word *panel* to refer to the rectangular area in SPSS Viewer in which a chart is placed. Normally the panels shrink automatically to fit; if you select this option, however, they remain full-size and wrap to the next line.

Building Graphs the Fast Way

The charts you build by choosing Graphs⇨Graphboard Template Chooser are the same as those you build by using the other menu selections, but you get less guidance along the way. So I suggest not building charts this way until you've used Chart Builder enough to know what chart-building choices do what you want done. Although the Graphboard option can be much faster — you just make quick selections and go — there is no clear signposts to follow to remind you where you are and what to do next.

First you select the variables you want to include, which causes the available kinds of charts (such as bar, dot, or line) to appear on-screen, as shown in Figure 9-11. Using the tabs at the top of the screen (Basic, Detailed, Titles, Options) you can choose screens that allow you to set the options. Be careful here; setting an option in one window may change the options that appear in another window. Be sure you know what you're after before you start. When you've set your preferred options, click the OK button, and the chart is generated.

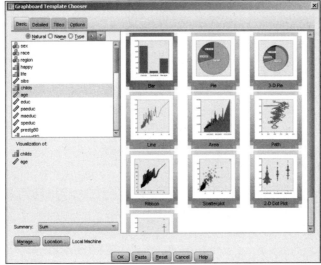

Figure 9-11:
The Interactive options for constructing a bar graph.

If you get things balled up, it's easy to restart. The Reset button removes everything you've entered in all the tabbed windows and restores all the defaults.

The Help button provides some information about whichever list of options is displayed at the moment.

The Paste button is for those who want to add to their graph definition in the most fundamental way possible. Inside SPSS, a graph is actually constructed by a command in the Syntax Command language. The steps you take to create a graph do nothing more than create the Command Syntax, which in turn creates the graph. The Paste button opens SPSS Syntax Editor with the Command Syntax in it, so you can edit the text of the command to produce the chart the way you want. (For more on using the Syntax language, see Chapters 14 and 15.)

Building Graphs the Old-Fashioned Way

The charts you build by choosing Graphs⇨Legacy Dialogs are simpler forms of the ones you build in the other ways, and the process is a bit different. As in the Graphboard process, you don't have the graphics and guidance you get from Chart Builder. The windows you use to set the variables are different from those in either of the other approaches. You don't have as many decisions to make as in the Graphboard, but you still need to be familiar with the process.

The first selection you make is the type of chart to be produced (such as bar, dot, or line). As you proceed through the steps that define the graph, different windows appear, such as the one shown for bar graphs in Figure 9-12. Each time you finish with one window, you click the OK button and move to the next window in the series. When you finish the last one, the result appears.

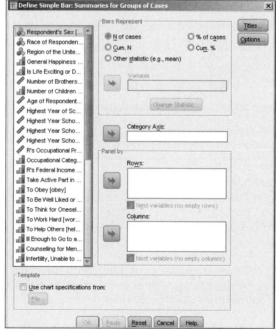

Figure 9-12: A window for designing a bar chart using the Legacy method.

The options presented to you in the Legacy method are not quite as complete as the ones in the other two methods. This arrangement makes it easy to produce simpler charts and graphs, but you must know what you're doing because you can't back up. When you've specified the values in a particular window and move on, the values stay that way until you've finished.

Editing a Graph

After you've built a chart and it's displayed in SPSS Viewer, you can still change it. Double-click the graph, and a copy of it appears in a new Chart Editor window, as shown in Figure 9-13.

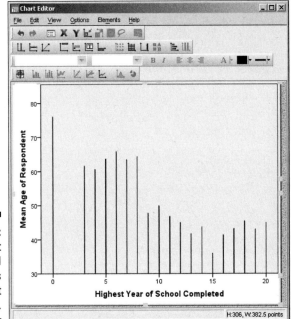

Figure 9-13:
A final chart presented for changes by Chart Editor.

Earlier in this chapter I mentioned that you could use templates to help define new charts in Chart Builder (see Figure 9-10). You can create a template file from Chart Editor by choosing File➪Save Chart Template and entering a filename.

Using Chart Editor, you can do a number of things with the chart. The options available are mostly the same ones you worked with when defining the original layout, so there are no big surprises.

Figure 9-14 is the same graph as Figure 9-13, but with the axes transposed (to make the bars grow horizontally), a fit line has been added, the overall size of the chart reduced, and a line has been added to emphasize a specific value.

The many menus of Chart Editor have option settings that you can use to try to make your chart demonstrate the data better. Fortunately, none of these selections are destructive — if you try something and don't like it, you can back your changes out and restore what you had before.

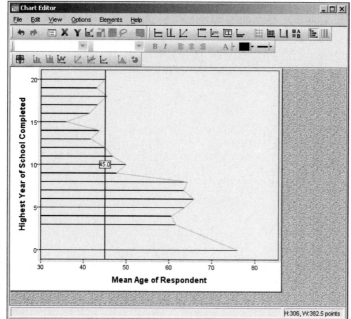

Chapter 10

Some Types of Graphs

. . . .

In This Chapter

▶ Drawing line charts with single and multiple lines

▶ Generating scatterplots

▶ Creating bar charts from your data

. . . .

This chapter provides examples of various graphical data displays and shows you how to build the graphs you're probably most familiar with; Chapter 11 shows examples of graph types that may be less familiar to you. Both chapters present each example as a step-by-step procedure, kept as simple as possible.

Although every variation of every possible chart won't fit into a pair of chapters, you can certainly use the procedures they present to produce some nifty-looking graphs. And once you get the basic idea of producing graphs, you should have no problem branching out and making fancy graphs of your own.

You could work through the examples in these two chapters to get an overview of building the kinds of graphs you can get from SPSS — not a bad idea for a beginner — or simply choose the look you want your data to have and follow the steps given here to construct the chart that does the job. Either way, when you get a handle on the basics, you can step through the process again and again, using your data, trying variations until you get charts that appear the way you want them to.

Line Chart

A *line chart* works well as a visual summary of categorical values. Line charts are also useful for displaying timelines because they demonstrate up and down trends so well. Line graphs are popular because they're easy to read. If they're not *the* most common type of statistical chart, they're a contender for the title.

TIP

The display of data is similar in a line chart and a bar chart. If you decide to display data as a line graph, you should probably try the same data as a bar chart to see which you prefer.

Simple line charts

The following steps generate a simple line chart displaying a single timeline:

1. **Choose File⇨Open⇨Data and open the** `Employee data.sav` **file, which is in the SPSS installation directory.**

2. **Choose Graphs⇨Chart Builder.**

   The Chart Builder dialog box appears.

3. **In the Choose From list, select Line.**

4. **Drag the first diagram (the one with the Simple Line tooltip) to the panel at the top.**

   An Element Properties dialog box appears. You can simply close it because this example uses the default settings.

5. **In the Variables list, drag Current Salary to the Y-Axis rectangle in the panel at the top.**

6. **Again in the Variables list, drag Date of Birth to the X-Axis rectangle in the panel.**

7. **Click the OK button.**

   The chart in Figure 10-1 appears.

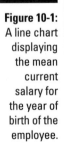

Figure 10-1:
A line chart displaying the mean current salary for the year of birth of the employee.

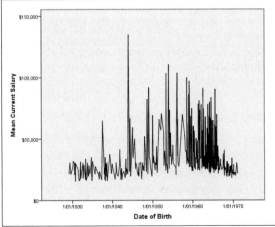

Charts with multiple lines

You can have more than one line appear on a chart by adding more than one variable name to an axis. But the variables must contain a similar range of values before they can be represented by the same axis. For example, if one variable ranges from 0 to 1,000 pounds and another variable ranges from 1 to 2 pounds, the values of the second variable will show up as a straight line, regardless of how much it actually fluctuates.

The following steps generate a multiline graph:

1. **Choose File⇨Open⇨Data and open the `Cars.sav` file.**

   The file is in the SPSS installation directory.

2. **Choose Graphs⇨Chart Builder.**

3. **In the Choose From list, select Line to specify the general type of graph to be constructed.**

4. **To specify that this graph should contain multiple lines, select the second diagram (the one with the Multiple Line tooltip) and drag it to the panel at the top.**

   The Element Properties dialog box pops up, but you can close it because the default values work fine.

5. **In the Variables list, select Number of Cylinders and drag it to the rectangle named X-Axis in the diagram.**

6. **In the Variables list, select Engine Displacement and drag it to the Y-Axis rectangle in the panel at the top.**

   The word *Mean* is added to the annotation because the values displayed on this axis will be the mean values of the engine displacement.

7. **In the Variables list, select Horsepower and drag it to the Y-axis also.**

   Be careful how you drop Horsepower. To add Horsepower as a new variable, you want to drop it on the little box containing the plus sign, as shown in Figure 10-2. If you drop the new name on top of the one that's already there, the original variable could be replaced.

Figure 10-2:
Adding another variable to the Y-axis.

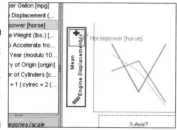

8. **When the Create Summary Group window appears, telling you that SPSS is combining the two variables along the Y-axis, click the OK button.**

9. **Click the OK button.**

   The chart shown in Figure 10-3 appears.

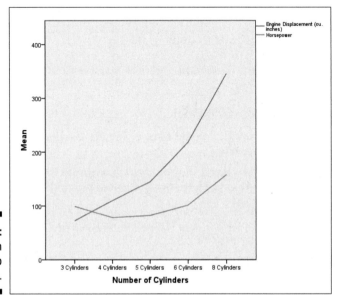

Figure 10-3:
A line graph
charting two
variables.

The variables you choose as members of the Y-axis must have a similar range of values to make sense. For example, if you were to choose age and annual income as two variables to be charted together, the result wouldn't be all that interesting; the salary values would be in the thousands and the ages, regardless of their variation, would all appear in a single line.

Scatterplots

A *scatterplot* is simply an X-Y plot where you don't care about interpolating the values — that is, the points are not joined with lines. Instead, a disconnected dot appears for each data point. The overall pattern of these scattered dots often exposes a pattern or trend.

Simple scatterplots

The following steps show you how to construct a simple scatterplot:

1. **Choose File⇨Open⇨Data and open the** `Employee data.sav` **file.**

   The file is in the SPSS installation directory.

2. **Choose Graphs⇨Chart Builder.**

3. **In the Choose From list, select Scatter/Dot.**

4. **Select the simplest scatterplot diagram (the one with the Simple Scatter tooltip), and drag it to the panel at the top.**

5. **In the Variables list, select Beginning Salary and drag it to the rectangle labeled X-Axis in the diagram.**

6. **In the Variables list, select Current Salary and drag it to the rectangle labeled Y-Axis in the diagram.**

7. **Click the OK button.**

   The chart in Figure 10-4 appears.

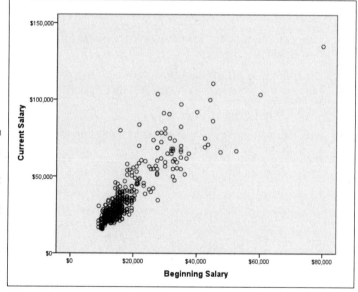

Figure 10-4:
A simple scatterplot showing the effect of starting salary on current salary.

Each dot on the scatterplot in Figure 10-4 represents both the starting salary and the current salary of one employee. The most obvious fact you can derive from this is that the current salary depends largely on the starting salary. In the pattern of the dots, it's easy to see a normal line from the lower left to the upper right. Any dot on that imaginary line represents the salary of an employee who received a normal raise. The dots above the line are the employees who got above-average raises, and those below the line are those with below-average raises. This plot has the shortcoming that the length of service is not considered.

Scatterplots with multiple variables

You can display the values of more than one variable along the same axis. The following example constructs a scatterplot showing the beginning salary and the current salary according to the number of months of experience the person had before taking the job:

1. **Choose File⇨Open⇨Data and open the** Employee data.sav **file, which is in the SPSS installation directory.**

2. **Choose Graphs⇨Chart Builder.**

3. **In the Choose From list, select Scatter/Dot.**

4. **Select the second scatterplot diagram (the one with the Grouped Scatter tooltip) and drag it to the panel at the top.**

5. **In the Variables list, do the following:**

    a. *Select Educational Level and drag it to the X-Axis rectangle.*

    b. *Select Beginning Salary and drag it to the Y-Axis rectangle.*

    c. *Select Current Salary and drag it to the same location as you dropped the Beginning Salary.*

    Be careful to drop it on the square with the plus sign. The plus sign appears as you drag a droppable item over the rectangle.

6. **Click the OK button.**

    The chart shown in Figure 10-5 appears, with two different colored dots and a legend at the upper right.

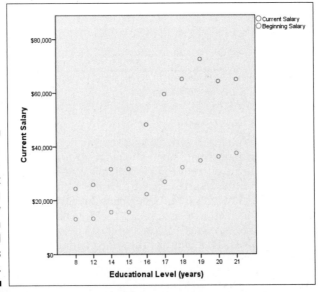

Figure 10-5: A scatterplot showing salary information organized by previous experience.

Simple three-dimensional scatterplots

Three-dimensional scatterplots can be dramatic in appearance, but clarity is not their strongest point. Because the scatterplot is drawn on a two-dimensional surface, you might find it difficult to envision where each point is supposed to appear in space. On the other hand, if your data distributes appropriately on the display, the resulting chart may demonstrate the concept you're trying to get across.

The following example uses the same data as in the preceding example but displays it in a different way, as a three-dimensional plot:

1. **Choose File⇨Open⇨Data and open the** Employee data.sav **file.**

   The file is in the SPSS installation directory.

2. **Choose Graphs⇨Chart Builder.**

3. **In the Choose From list, select Scatter/Dot.**

4. **Select the third scatterplot diagram (the one with the Simple 3-D Scatter tooltip) and drag it to the panel at the top.**

5. **In the Variables list, do the following:**

   a. *Select Beginning Salary and drag it to the X-Axis rectangle.*

   b. *Select Current Salary and drag it to the Y-Axis rectangle.*

   c. *Select Previous Experience and drag it to the Z-Axis rectangle.*

6. **Click the OK button.**

   The graph shown in Figure 10-6 appears.

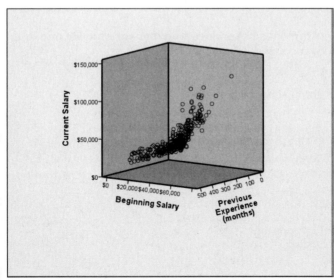

Figure 10-6:
A three-dimensional scatterplot printed on a two-dimensional surface.

Grouped three-dimensional scatterplots

Three-dimensional scatterplots can have more than a single value along an axis. However, as in the previous example, clarity is not the strongest point of the result. Depending on the data, you might find it difficult to envision where each point is supposed to appear in space — or the chart may demonstrate the concept you're trying to get across. All you can do is generate the graph and then evaluate whether or not you want to use the result.

The following example uses the same data basic information that was displayed in the preceding example:

1. **Choose File⇨Open⇨Data and open the** `Employee data.sav` **file.**

   The file is in the SPSS installation directory.

2. **Choose Graphs⇨Chart Builder.**

3. **In the Choose From list, select Scatter/Dot.**

4. **Select the fourth scatterplot diagram (the one with the Grouped 3-D Scatter tooltip) and drag it to the panel at the top.**

5. **In the Variables list, do the following:**

   a. *Select Educational Level and drag it to the Z-Axis rectangle.*

   b. *Select Current Salary and drag it to the Y-Axis rectangle.*

   c. *Select Beginning Salary and drag it to the plus sign in the Y-Axis rectangle.*

6. **Drag INDEX from the X-Axis rectangle to the Set Color rectangle at the upper right.**

   The INDEX variable is automatically inserted by the graph-building process. It is not used in the example.

7. **Select Previous Experience from the Variable list and drag it to the X-Axis rectangle.**

8. **Click the OK button.**

   The graph shown in Figure 10-7 appears.

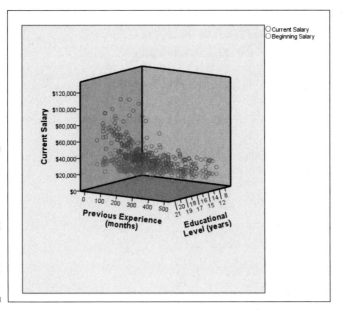

Figure 10-7:
A grouped three-dimensional scatterplot printed on a two-dimensional surface.

Summary Point plots

A Summary Point plot has the same layout as a bar chart, except the bars are not drawn. Instead, a point is placed at the position of the top of the bar. This means that it is possible to place more than a single value at the position that would otherwise be filled by a single bar. The points for different data items vary in appearance, and are identified by a legend. You must be careful, however, to choose variables that make sense when combined and displayed this way.

The following example shows the relationship between the means of cubic-inch engine displacement and the mean of automobile horsepower over several years:

1. **Choose File⇨Open⇨Data and open the** `Cars.sav` **file.**

   The file is in the SPSS installation directory.

2. **Choose Graphs⇨Chart Builder.**

3. **In the Choose From list, select Scatter/Dot.**

4. **Select the fifth scatterplot diagram (the one with the Summary Point plot tooltip) and drag it to the panel at the top.**

5. **In the Variables list, do the following:**

    a. *Select Model Year and drag it to the X-Axis rectangle.*

    b. *Select Horsepower and drag it to the Y-Axis rectangle.*

    c. *Select Engine Displacement and drag it to the plus sign in the Y-Axis rectangle.*

6. **Click the OK button.**

    The graph shown in Figure 10-8 appears.

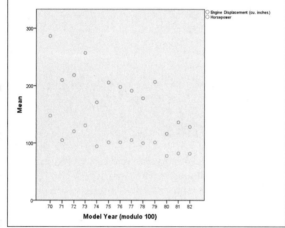

Figure 10-8:
A Summary
Point plot
showing the
difference
of the trends
of two
variables.

Simple Dot plots

No plot is simpler to produce than the *dot plot*. It has only one dimension. Although SPSS groups it among the scatterplots, there's nothing scattered about it. It actually presents data more like a bar chart — and it reminds me of that old joke about stacking BBs.

It's easy to create a dot plot. You select the dot plot as the type of graph you want and then select one variable. SPSS does the rest. The following steps guide you through the process of creating a simple dot plot:

1. **Choose File⇨Open⇨Data and open the** `Employee data.sav` **file.**

    The file is in the SPSS installation directory.

2. **Choose Graphs⇨Chart Builder.**

3. **In the Choose From list, select Scatter/Dot.**

4. **Select the sixth graph image (the one with the Simple Dot Plot tooltip) and drag it to the panel at the top.**

5. **In the Variables list, select Date of Birth and drag it to the X-Axis rectangle.**

6. **Click the OK button.**

   The chart shown in Figure 10-9 appears.

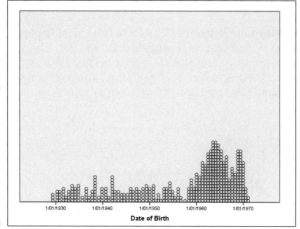

Figure 10-9:
A dot plot showing the relative numbers of persons born in each year.

Scatterplot matrices

A *scatterplot matrix* is a group of scatterplots combined into a single graphic image. You choose a number of scale variables and include them as a member of your matrix, and SPSS creates a scatterplot for each possible pair of variables. You can make the matrix as large as you like — its size is controlled by the number of variables you include.

The following steps walk you through the creation of a matrix:

1. **Choose File⇨Open⇨Data and open the** `Employee Data.sav` **file.**

   The file is in the SPSS installation directory.

2. **Choose Graphs⇨Chart Builder.**

3. **In the Choose From list, select Scatter/Dot.**

4. **Select the seventh graph image (the one with the Scatterplot Matrix tooltip) and drag it to the panel at the top.**

5. **In the Variables list, drag Beginning Salary to the Scattermatrix rectangle in the panel at the top.**

   The selected name replaces the label in the rectangle.

6. **Drag the variable names Current Salary, Months Since Hire, and Previous Experience (Months) to the rectangle inside the panel at the top of the window.**

   The labels may or may not change with each variable you add, depending on their length and amount of space available. All your labels appear in the list at the bottom of the Element Properties dialog box.

7. **Click the OK button.**

   The chart in Figure 10-10 appears. As you can see, each variable is plotted against each of the others.

The matrix of scatterplots in Figure 10-10 has each variable plotted against each of the others. Notice that the scatterplots along the diagonal from the upper left to the lower right are blank — that's because it's useless to plot a variable against itself. Also, notice the symmetry: Each plot in the lower-left half has a rotated and mirrored image in the upper-right half.

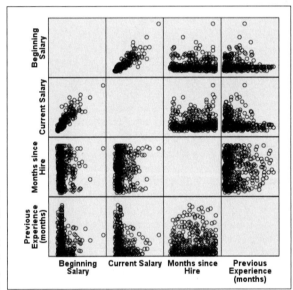

Figure 10-10: A four-by-four matrix of scatter-plots.

Drop-line charts

A *drop-line chart* presents a special kind of summary with points and vertical lines. The points are grouped horizontally; at each categorical value, a line is drawn vertically through them. This arrangement can be visually helpful for comparing the values that appear within each category.

The following steps take you through the basic actions necessary for producing a drop-line graph:

1. **Choose File⇨Open⇨Data and open the** `Cars.sav` **file.**

   The file is in the SPSS installation directory.

2. **Choose Graphs⇨Chart Builder.**

3. **In the Choose From list, select Scatter/Dot.**

4. **Select the last graph image (the one with the Drop-line tooltip) and drag it to the panel at the top.**

5. **In the Variables list, do the following:**

   a. *Select Number of Cylinders and drag it to the rectangle in the upper-right corner with the Set Color label.*

   b. *Select Model Year and drag it to the X-Axis rectangle.*

   c. *Select Horsepower and drag it to the rectangle on the left labeled Mean.*

   Note that X-Axis and Set Color both contain categorical variable names, and the Mean on the left contains a scale variable. This is the only combination of variable types that will work.

6. **Click the OK button.**

   The graph in Figure 10-11 appears.

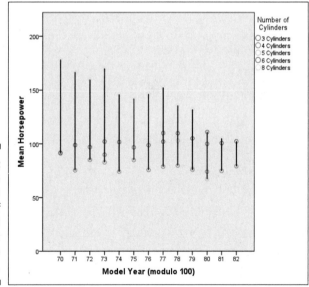

Figure 10-11: A drop-line plot is based on a pair of categorical values and a scale variable.

Bar Graphs

A *bar graph* is a comparison of relative magnitudes. Simple bar graphs and simple line graphs are the most common ways of charting statistics. It would make an interesting statistical study to determine which is more common. The results could be displayed as either a bar graph or a line graph, whichever is more popular.

Simple bar graphs

A fundamental bar graph is simple enough that the decisions you need to make when preparing one are almost intuitive. The following steps can be used to generate a simple bar graph:

1. **Select File⇨Open⇨Data and open the** `Employee data.sav` **file.**

   The file is in the SPSS installation directory.

2. **Choose Graphs⇨Chart Builder.**

3. **In the Choose From list, select Bar.**

4. **Select the first graph image (the one with the Simple Bar tooltip) and drag it to the panel at the top of the window.**

5. **In the Variables list, select Education Level and drag it to the X-Axis rectangle.**

6. **In the Variables list, select Current Salary and drag it to the Count rectangle.**

   The label changes from Y-Axis to Mean to indicate the type of variable that will now be applied to that axis.

7. **Click the OK button.**

   The bar graph in Figure 10-12 appears.

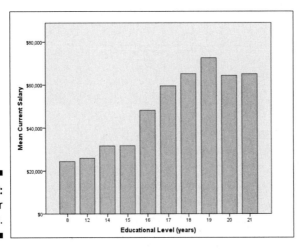

Figure 10-12:
A simple bar graph.

Clustered bar charts

A *clustered bar chart* can show the relationships among a cluster of items by displaying more than one value and presenting a summary of categorical values. Clustering combines several bar charts into one. The following steps take you through the process of constructing a clustered bar chart:

1. **Choose File⇨Open⇨Data and open the** `Cars.sav` **file.**

   The file is in the SPSS installation directory.

2. **Choose Graphs⇨Chart Builder.**

3. **In the Choose From list, select Bar.**

4. **Select the second graph image (the one with the Clustered Bar tooltip) and drag it to the panel at the top of the window.**

5. **In the Variables list, do the following:**

   a. **Select Model Year and drag it to the X-Axis rectangle.**

   b. **Select Horsepower and drag it to the Count rectangle.**

   The rectangle was originally labeled Y-Axis. The label changed to help you understand the type of variable that needs to be placed there.

   c. **Select Number of Cylinders and drag it to the rectangle in the upper-right corner, the one now labeled Cluster on X.**

6. **Click the OK button.**

   The graph in Figure 10-13 appears.

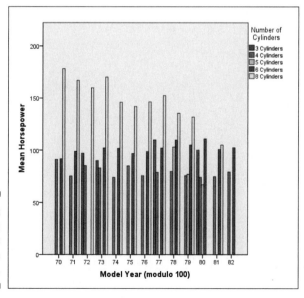

Figure 10-13: A bar graph with values displayed as clusters.

Stacked bar charts

A *stacked bar chart* is similar to the clustered bar chart in that it displays multiple values of a variable for each value of a categorical variable. But it does it by stacking them instead of placing them side by side. The following chart displays the same data as the preceding example, but emphasizes different aspects of the data.

Following these steps will create a stacked bar chart:

1. **Choose File⇨Open⇨Data and open the** Cars.sav **file.**

   The file is in the SPSS installation directory.

2. **Choose Graphs⇨Chart Builder.**

3. **In the Choose From list, select Bar.**

4. **Select the third graph image (the one with the Stacked Bar tooltip) and drag it to the panel at the top of the window.**

5. **In the Variables list, do the following:**

   a. *Select Model Year and drag it to the X-Axis rectangle.*

   b. *Select Horsepower and drag it to the Count rectangle.*

      The rectangle was originally labeled Y-Axis. The label changed to help you understand the type of variable that needs to be placed there.

   c. *Select Number of Cylinders and drag it to the rectangle in the upper-right corner, the one now labeled Stack.*

6. **Click the OK button.**

   The graph in Figure 10-14 appears.

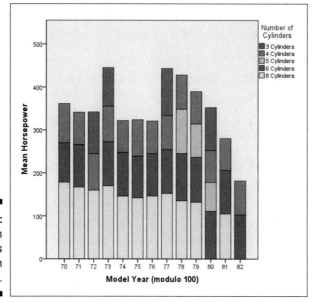

Figure 10-14:
A bar graph
with values
displayed in
stacks.

Simple three-dimensional bar charts

A *simple three-dimensional bar chart* is the same as a two-dimensional bar chart, except a third variable is added to specify the values along the new dimension. As with most three-dimensional displays, it has the advantage of displaying three relative values at once, and it has the disadvantage of making it difficult to determine which is the greater of two values if the two values are close. You'll just have to try it with your data and decide whether it demonstrates what you've got or just contributes to the general confusion.

The following steps construct a three-dimensional bar chart:

1. **Choose File⇨Open⇨Data and open the** Cars.sav **file.**

   The file is in the SPSS installation directory.

2. **Choose Graphs⇨Chart Builder.**

3. **In the Choose From list, select Bar.**

4. **Select the fourth graph image (the one with the Simple 3-D Bar tooltip) and drag it to the panel at the top of the window.**

5. **In the Variable list, do the following:**

   a. Select Model Year and drag it to the Y-Axis rectangle.

   b. Select Number of Cylinders and drag it to the X-Axis rectangle.

   c. Select Country of Origin and drag it to the Z-Axis rectangle.

6. **Click the OK button.**

   The graph in Figure 10-15 appears.

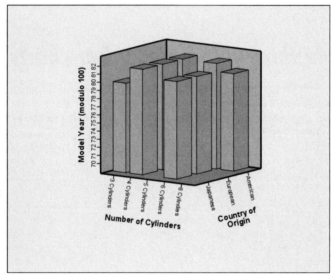

Figure 10-15:
A three-dimensional bar graph.

Clustered three-dimensional bar charts

A *clustered three-dimensional bar chart* can show the relationships among a cluster of items by displaying more than one value and presenting a summary of categorical values. Clustering combines several bar charts into one. A great deal of information is displayed in a single chart, and it can be confusing in appearance. You will need to test it with your data and determine whether it demonstrates what you are trying to show.

The following steps take you through the process of constructing a clustered bar chart:

1. **Choose File⇨Open⇨Data and open the** `Cars.sav` **file.**

   The file is in the SPSS installation directory.

2. **Choose Graphs⇨Chart Builder.**

3. **In the Choose From list, select Bar.**

4. **Select the fifth graph image (the one with the Clustered 3-D Bar tooltip) and drag it to the panel at the top of the window.**

5. **In the Variables list, do the following:**

   a. *Select Number of Cylinders and drag it to the X-Axis rectangle.*

   b. *Select Horsepower and drag it to the Count rectangle.*

   The rectangle was originally labeled Y-Axis. The label changed to help you understand the type of variable that needs to be placed there.

   c. *Select Country of Origin and drag it to the Z-Axis rectangle.*

   d. *Select Model Year and drag it to the rectangle in the upper-right corner, the one now labeled Cluster on X.*

6. **Click the OK button.**

   The graph in Figure 10-16 appears.

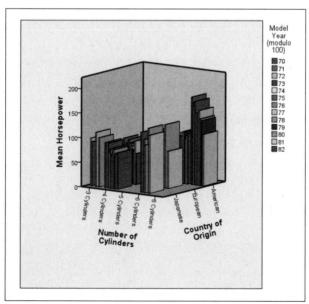

Figure 10-16:
A bar graph with values displayed as clusters.

Stacked three-dimensional bar charts

A *stacked three-dimensional bar chart* is very similar to the clustered bar chart. The two can be used to show the same data, but in slightly different ways. A stacked chart can show the relationships among data items by stacking more than one value and presenting a vertical summary of categorical values. Stacking combines several bar charts into one. Because that single chart displays a great deal of information, however, it can be confusing in appearance. The layout may even hide some of the data inadvertently. You'll have to test it with your data to see whether it shows what you want to see.

The following steps take you through the process of constructing a stacked three-dimensional bar chart:

1. **Choose File⇨Open⇨Data and open the** `Cars.sav` **file.**

   The file is in the SPSS installation directory.

2. **Choose Graphs⇨Chart Builder.**

3. **In the Choose From list, select Bar.**

4. **Select the sixth graph image (the one with the Stacked 3-D Bar tooltip) and drag it to the panel at the top of the window.**

5. **In the Variables list, do the following:**

   a. Select Number of Cylinders and drag it to the X-Axis rectangle.

   b. Select Horsepower and drag it to the Count rectangle.

   The rectangle was originally labeled Y-Axis. The label changed to help you understand the type of variable that should be placed there.

   c. Select Country of Origin and drag it to the Z-Axis rectangle.

   d. Select Model Year and drag it to the rectangle in the upper-right corner, the one labeled Stack Set Color.

6. **Click the OK button.**

   The graph in Figure 10-17 appears.

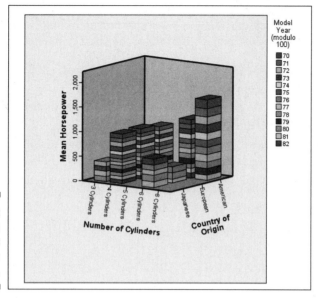

Figure 10-17:
A bar graph
with values
displayed in
stacks.

Simple error bars

Some errors come from flat-out mistakes — but those aren't the errors I talk about here. Statistical sampling can help you arrive at a conclusion, but that conclusion has a margin of error. This margin can be calculated and quantified according to the size of the sample and the distribution of the data. For example, suppose you want to know how typical the result is when you calculate the mean of all values for a particular variable — for any one case, the mean could be as big as the largest value or as small as the smallest. The maximum and minimum are the extremes of the possible error. You can choose values and mark the points that contain, say, 90 percent of all values. Marking these points on graphs creates *error bars*.

You can add error bars to the display of most types of graphs. For example, you could add error bars to the simple bar graph presented earlier in this chapter (refer to Figure 10-12) by making selections in the Element Properties dialog box. If you've worked through any of the examples, you'll know Element Properties as that pesky window that pops up every time you construct a chart.

For an example of adding error bars to a bar chart, follow the same procedure described previously in the "Simple bar graphs" section — but just before the final step (clicking the OK button to produce the chart), do the following:

1. **If the Element Properties window is not displayed, click the Element Properties button to put that window on-screen.**

2. **In the Element Properties window, make sure that a check mark appears in the Display Error Bars option.**

3. **Select Confidence Level Intervals and set its value to 95%.**

4. **Click the Apply button.**

5. **Close the Element Properties window by selecting the Close button.**

6. **Click the OK button.**

   The chart shown in Figure 10-18 appears.

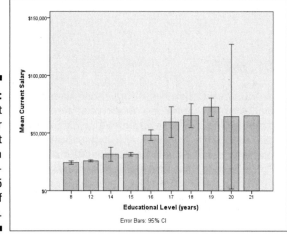

Figure 10-18: A bar chart with error bars that show a range containing 95 percent of all values.

Another way to display the same data is to display the range of errors without displaying the full range of all values. To do so, follow these steps:

1. **Choose File⇨Open⇨Data and open the** Employee data.sav **file.**

   The file is in the SPSS installation directory.

2. **Choose Graphs⇨Chart Builder.**

3. **In the Choose From list, select Bar.**

4. **Select the seventh graph image (the one with the Simple Error Bar tooltip) and drag it to the panel at the top of the window.**

5. **In the Element Properties window, make sure that the Display Error Bars option is checked, the Confidence Intervals option is selected, and the Level is set to 95%.**

Any time you change a setting or a value in the Element Properties dialog box, you must click the Apply button to have the change reflected in your chart.

6. **In the Variables list, select Education Level and drag it to the X-Axis rectangle.**

7. **In the Variables list, select Current Salary and drag it to the Mean rectangle.**

The label changes from Y-Axis to Mean to indicate the type of data that will be displayed on that axis.

8. **Click the OK button.**

The bar graph in Figure 10-19 appears.

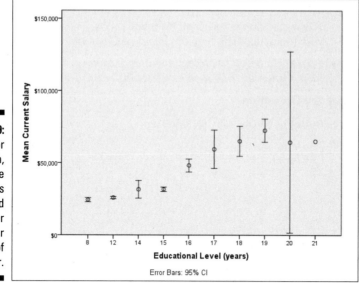

Figure 10-19: An error bar graph, showing the mean values as dots and the upper and lower bounds of the error.

This example displays the result of one way of making error calculations: The magnitude of the error is based on 95 percent of all values being within the upper and lower error bounds. If you prefer, you can base the error on the bell curve and mark the upper and lower errors at some multiple of the standard error or standard deviation.

Clustered error bars

A *clustered error bar chart* can show the relationships among a cluster of error ranges by presenting a summary of categorical values. Clustering combines several error bar charts into one. The following steps take you through the process of constructing a clustered error bar chart:

1. **Choose File⇨Open⇨Data and open the** Cars.sav **file.**

   The file is in the SPSS installation directory.

2. **Choose Graphs⇨Chart Builder.**

3. **In the Choose From list, select Bar.**

4. **Select the eighth graph image (the one with the Clustered Error Bar tooltip) and drag it to the panel at the top of the window.**

5. **In the Variables list, do the following:**

   a. Select Model Year and drag it to the X-Axis rectangle.

   b. Select Horsepower and drag it to the Mean rectangle.

   The rectangle was originally labeled Y-Axis. The label changed to help you understand the type of variable that should be placed there.

   c. Select Number of Cylinders and drag it to the rectangle in the upper-right corner, the one now labeled Cluster on X.

6. **Click the OK button.**

   The graph in Figure 10-20 appears.

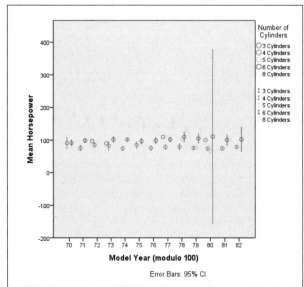

Figure 10-20:
An error bar graph, showing error ranges displayed as clusters.

Chapter 11

More Types of Graphs

In This Chapter

▶ Displaying histograms and area graphs

▶ Making use of pie charts and three kinds of boxplots

▶ Using dual-axis charts to combine variables with different ranges

SPSS has a number of ways to present data graphically; depending on the characteristics of your data, some graphs are more appropriate than others. Chapter 10 provides an overview of some more common graph types; this chapter discusses some lesser-known ones. Every example in these two chapters is as simple as possible to present a general idea of the types of charts you can choose. To use one with your data, you start by choosing a basic form and then continue by setting options to extend the form so it displays the information in the way that works best for your purposes. The Element Properties window, which appears automatically every time you generate a new chart, provides every possible option that applies to the chart you're building.

When you use Chart Builder, it's completely safe to drag and drop any variables you want to see in your graph; if the variable won't make sense there, the drop will fail. SPSS does you the kindness of figuring out what will and won't work. Also, no matter what you try to do while building a graph, your data will never be hurt.

Histograms

A *histogram* represents the number of items that appear within a range of values (or within a bin, statistically speaking — see Chapter 7). You can use a histogram to look at a graphic representation of the frequency distribution of the values of a variable. Histograms are useful for demonstrating the patterns in your data when you want to display information to others rather than discover data patterns for yourself.

Simple histograms

You can use the following steps to create a simple histogram that displays the number of automobiles (in the survey used in the example) that have particular gas mileage capabilities for each of several years:

1. **Choose File⇨Open⇨Data and open the** `Cars.sav` **file, which is in the SPSS installation directory.**

2. **Choose Graphs⇨Chart Builder.**

    The Chart Builder dialog box appears.

3. **In the Choose From list, select Histogram.**

4. **Drag the first graph diagram (the one with Simple Histogram tooltip) to the panel at the top of the window.**

5. **In the Variables list, do the following:**

    a. Select the Model Year variable and drag it to the X-Axis rectangle in the panel.

    b. Select Miles Per Gallon and drag it to the Count rectangle on the left side of the panel.

6. **Click the OK button.**

    The histogram shown in Figure 11-1 appears.

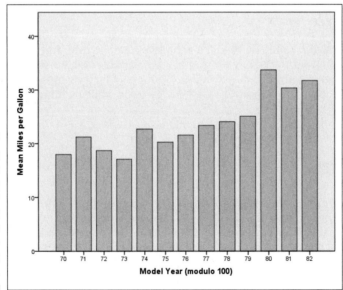

Figure 11-1:
A histogram displaying the number of cars with various gas mileage values in each year.

The graph in Figure 11-1 looks like a bar chart, but it isn't. The height of each bar does not represent the mean or an average — the height is determined by the largest value.

The meaning of a graph of this sort is not intuitive; you may want to add a note explaining what it means.

Stacked histograms

You can create a histogram that is more like a bar chart. In a *stacked histogram,* the overall extent of the bars represents the sum of the values in each category, and different categories of a third variable are indicated by displaying portions of the bars in different colors.

In this type of histogram, the scale on the left can be used to gauge the relative sizes of each of the colored segments of a bar. The overall height of each bar is the sum of the miles per gallon in each model year (as it would be in a bar chart). Here each bar is a stack of rectangles, each one representing a portion of the total number of cars — in this case, cars with a certain number of cylinders. The following steps produce a stacked histogram displaying the same information as shown in the preceding simple histogram, but this one displays sums instead of means:

1. **Choose File⇨Open⇨Data and open the `Cars.sav` file.**

2. **Choose Graphs⇨Chart Builder.**

3. **In the Choose From list, select Histogram.**

4. **Drag the second graph diagram (the one with the Stacked Histogram tooltip) to the panel at the top of the window.**

5. **In the Variables list, do the following:**

    a. *Select the Model Year variable and drag it to the X-Axis rectangle.*

    b. *Select Miles Per Gallon and drag it to the Count rectangle on the left side of the panel.*

    c. *Select Number of Cylinders and drag it to the Stack Set Color rectangle, at the upper right.*

6. **Click the OK button.**

    The histogram shown in Figure 11-2 appears.

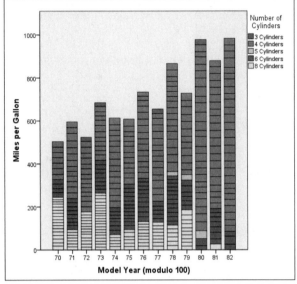

Figure 11-2:
A stacked
histogram,
highlighting
the number
of cars with
specific
numbers of
cylinders.

Frequency polygons

A *frequency polygon* is a histogram that looks like an area graph (described in the next section) — but it is a histogram, so it works differently. It represents the number of items that appear within a range. The following steps guide you through a procedure that produces a frequency polygon histogram:

1. **Choose File⇨Open⇨Data and open the** `Cars.sav` **file.**

2. **Choose Graphs⇨Chart Builder.**

3. **In the Choose From list, select Histogram.**

4. **Drag the third graph diagram (the one with the Frequency Polygon tooltip) to the panel at the top of the window.**

5. **In the Variables list, do the following:**

   a. *Select the Model Year variable and drag it to the X-Axis rectangle in the panel.*

   a. *Select Miles Per Gallon and drag it to the Mean rectangle.*

   This is the rectangle that was originally called the Y-Axis.

6. **Click the OK button.**

   The histogram shown in Figure 11-3 appears.

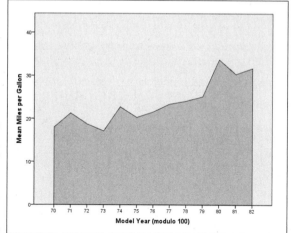

Population pyramids

A *population pyramid* provides an immediate comparison of the number of
items that fall into categories. It is called a pyramid because it often takes
a triangular shape — wide at the bottom and tapering to a point at the top.
The following steps can be followed to build an example pyramid histogram
chart:

1. **Choose File⇨Open⇨Data and open the** `Employee data.sav` **file,
   which is in the SPSS installation directory.**

2. **Use the tab to switch to Variable View.**

3. **Select the Type column of the** `bdate` **variable.**

4. **Click the button that appears near the variable type name, which is**
   `Date.`

5. **In the list of date formats, choose** `mmm yy` **and then click the OK
   button.**

   This is a matter of personal preference. The chart is produced no matter
   which format is used to display the dates, but I think this format looks
   better than most of the others.

6. **Choose Graphs⇨Chart Builder.**

7. **In the Choose From list, select Histogram.**

8. **Drag the fourth graph diagram (the one with the Population Pyramid
   tooltip) to the panel at the top of the window.**

9. **In the Variables list, do the following:**

   a. *Select the Gender variable and drag it to the Split Variable rectangle.*

   This is a categorical variable with two possible values, so one category will be placed on each side of the center line.

   b. *Select Date of Birth and drag it to the Distribution Variable rectangle.*

10. **Click the OK button.**

   The chart shown in Figure 11-4 appears.

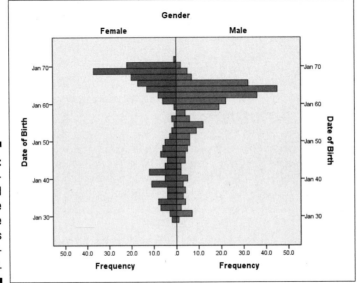

Figure 11-4: A population pyramid shows the occurrence of values within categories.

You can create pyramid histograms based on categorical variables with three, four, or more values. The plot produced will consist of as many pairs as needed (and even a single-sided pyramid for one category, if necessary) to display bars that show the relative number of occurrences of different values in the categories.

Area Graphs

An *area graph* is really a line graph, or a collection of line graphs, with the areas below the lines filled in to represent the mean of one or more values at the various points.

Simple area graphs

A *simple area graph* displays the area below a single line. The following steps produce a simple area graph:

1. **Choose File⇨Open⇨Data and open the** `Employee data.sav` **file, which is in the SPSS installation directory.**

2. **Choose Graphs⇨Chart Builder.**

3. **In the Choose From list, select Area.**

4. **Drag the first graph diagram (the one with the Simple Area tooltip) to the panel at the top of the window.**

5. **In the Variables list, do the following:**

   a. *Select the Educational Level variable and drag it to the X-Axis rectangle.*

   b. *Select Beginning Salary and drag it to the Count rectangle.*

   This is the rectangle that was labeled Y-Axis until the X-Axis became defined.

6. **Click the OK button.**

   The area chart shown in Figure 11-5 appears.

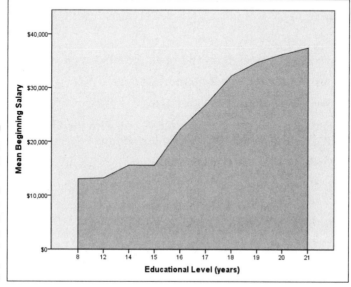

Figure 11-5: An area chart showing the mean starting salary for various levels of education.

Stacked area charts

A *stacked area chart* is a chart with more than one variable being displayed along the X-axis. The values are stacked in such a way that the ups and downs of the lower value in the chart affect the upper values in the chart. That is, the chart is not a group of independent lines; instead, it represents a cumulative total — to which each variable displayed adds a value.

Step 5c of the following procedure can be repeated for the inclusion of two or more variables. They all appear in the legend at the upper right, and each variable provides the value for one layer of the stack.

If you include more than one variable, make sure that the variables you select for stacking have similar ranges of values so the scale on the left side will make sense for all of them. If, for example, one variable ranges into the thousands and the other doesn't go over a hundred, the smaller one will compress itself visually — and come out in the final graph as a line.

When you select multiple variables for stacking, be sure to select them in the order you want them stacked; the first one you select will remain on top. The second one you select will be placed under it, and so on.

The two types of area charts, simple and stacked, act the same when you construct them. You can select the stacked chart and produce a single-area chart, or you can start with the simple area chart and stack your variables.

Follow these steps to produce a stacked area chart with two stacked variables:

1. **Choose File➪Open➪Data and open the `Employee data.sav` file.**

2. **Choose Graphs➪Chart Builder.**

3. **In the Choose From list, select Area.**

4. **Drag the second graph diagram (the one with the Stacked Area tooltip) to the panel at the top of the window.**

5. **In the Variables list, do the following:**

    a. *Select the Educational Level variable and drag it to the X-Axis rectangle.*

    b. *Select Current Salary and drag it to the Count rectangle.*

       This is the rectangle that was originally called the Y-Axis.

    c. *Select Beginning Salary and drag it to the plus sign in the Current Salary rectangle.*

       Be sure to drag it to the plus sign and not simply to the rectangle in general. (The plus sign appears at the top of the rectangle when you drag the new variable's name across it.)

6. **Click the OK button.**

The area chart shown in Figure 11-6 appears.

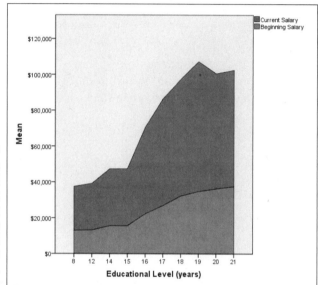

Figure 11-6:
A stacked
area chart
showing
one variable
being added
to another in
the display.

Pie Charts

Pie charts are the easiest kind to spot — they are the only charts that show up as circles. The purpose of a *pie chart* is simply to show how something (the "whole") is divided into pieces — whether two, ten, or any other number. Each slice in the pie chart represents its percentage of the whole. For example, if a slice takes up 40 percent of the total pie, that slice represents 40 percent of the total number. A pie chart is also called a *polar chart*, so SPSS calls this option Pie/Polar.

In the following steps, you construct a simple pie chart:

1. **Choose File⇨Open⇨Data and open the** `Employee data.sav` **file, which is in the SPSS installation directory.**

2. **Choose Graphs⇨Chart Builder.**

3. **In the Choose From list, select Pie/Polar.**

4. **Drag the pie diagram to the panel at the top of the window.**

5. **In the Variables list, drag Educational Level to the Slice By rectangle at the bottom of the panel.**

6. **Click the OK button.**

The pie chart shown in Figure 11-7 appears.

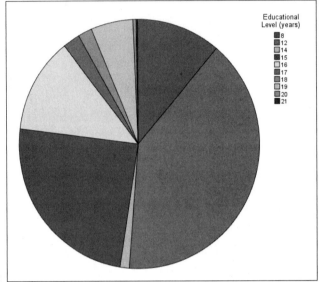

Educational
Level (years)
- 8
- 12
- 14
- 15
- 16
- 17
- 18
- 19
- 20
- 21

Figure 11-7:
A pie chart
displaying
the number
of employ-
ees at each
education
level.

Boxplots

A *boxplot* uses graphic elements to display five statistics at one time within each categorical value. The statistics are the minimum value, first quartile, median value, third quartile, and maximum value. A boxplot is particularly good for helping you spot values lying well outside the range of normal values.

Simple boxplots

A simple boxplot displays the range of values of a single scale variable for all values of a categorical variable. The following steps guide you through the creation of a simple boxplot:

1. **Choose File⇨Open⇨Data and open the** `Employee data.sav` **file, which is in the SPSS installation directory.**

2. **Choose Graphs⇨Chart Builder.**

3. **In the Choose From list, select Boxplot.**

4. **Drag the first graph diagram (the one with the Simple Boxplot tooltip) to the panel at the top of the window.**

5. **In the Variables list, do the following:**

    a. Select the Educational Level variable and drag it to the X-Axis rectangle.

    b. Select the Current Salary variable and drag it to the Y-Axis rectangle.

6. **Click the OK button.**

    The boxplot shown in Figure 11-8 appears.

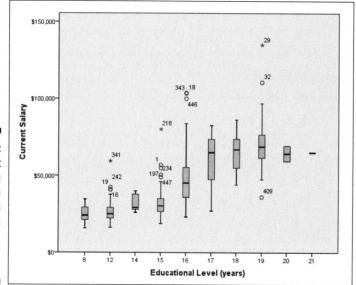

Figure 11-8:
A boxplot
displaying
the range
of values
for each
value of a
categorical
variable.

In Figure 11-8, each vertical column of graphics represents all the values for a category. Values beyond the extents of the first and third quartiles are marked with circles or stars; those marked with stars represent extremes. You can look at a boxplot of this type to find where your data may be out of whack.

Clustered boxplots

A *clustered boxplot* displays the values of three variables in one graph. A boxplot displays a lot of information — and if it's displaying three variables, it can get very busy visually. Fortunately, it's easier to read on-screen in color than it is here on this page in shades of gray. The legend in the upper-right corner assigns colors to the categorical values; those colors appear in the boxes to show you which is which. You're also shown the ID numbers of cases with extreme values. Use the following steps to construct a clustered boxplot:

1. **Choose File⇨Open⇨Data and open the** `Employee data.sav` **file.**

2. **Choose Graphs⇨Chart Builder.**

3. **In the Choose From list, select Boxplot.**

4. **Drag the second graph diagram (the one with the Clustered Boxplot tooltip) to the panel at the top of the window.**

5. **In the Variables list, do the following:**

   a. *Drag the Minority Classification variable to the X-Axis rectangle.*

   b. *Drag the Current Salary variable to the Y-Axis rectangle.*

   c. *Drag the Educational Level variable to the Cluster on X rectangle.*

6. **Click the OK button.**

   The boxplot shown in Figure 11-9 appears.

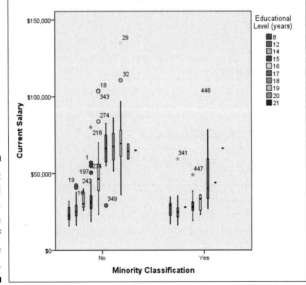

Figure 11-9:
A clustered boxplot displaying the values of three variables.

One-dimensional boxplots

A *one-dimensional boxplot* displays one variable in such a way that you can easily see the range of values and spot out-of-range values. The following steps construct an example of a one-dimensional boxplot:

1. **Choose File⇨Open⇨Data and open the** `Employee data.sav` **file.**

2. **Choose Graphs⇨Chart Builder.**

3. **In the Choose From list, select Boxplot.**

4. **Drag the third graph diagram (the one with the 1-D Boxplot tooltip) to the panel at the top of the window.**

5. **Click the Groups/Point ID tab and select the Point To ID Label option.**

   A rectangle labeled Point Label Variable appears in the upper-right corner of the preview panel at the top.

6. **In the Variables list, do the following:**

   a. *Drag the Employee Code variable to the Point Label rectangle in the upper-right of the panel.*

   b. *Drag the Current Salary variable to the X-Axis rectangle, on the left side of the panel.*

7. **Click the OK button.**

   The boxplot shown in Figure 11-10 appears.

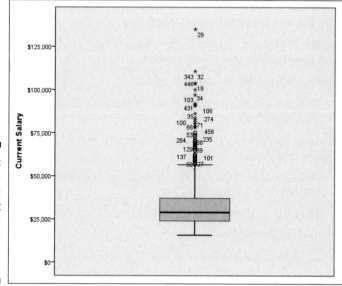

Figure 11-10: A one-dimensional boxplot showing extreme values of a variable.

The boxplot in Figure 11-10 graphically displays values out of the normal range. Each value is tagged with the ID number of its case. The number displayed as the ID is the variable previously chosen as the point ID. If no point ID variable had been chosen, the annotation would show the normal SPSS case numbers.

High-Low Graphs

A *high-low chart* displays the range of values between specified high and low values. Its purpose is to compare two or three variables.

High-low-close graphs

The *high-low-close graph* shows how a variable appears when plotted between a high value and a low value. The graph displays the relationships among three sets of values. This example and the one that follows (the *simple-range bar graph*) display the same information, but with a different look to the graphics. This graph displays the range as a vertical line, and the other graph shows it as a bar.

Follow these steps to create a high-low-close graph:

1. **Choose File➪Open➪Data and open the file named** Home sales [by neighborhood].sav, **which is in the SPSS installation directory.**

2. **Choose Graphs➪Chart Builder.**

3. **In the Choose From list, select High-Low.**

4. **Drag the first graph diagram (the one with the High-Low-Close tooltip) to the panel at the top of the window.**

5. **In the Variables list, do the following:**

    a. *Drag the Total Appraised Value variable to the High Variable rectangle at the top on the left.*

    b. *Drag the Appraised Land Value variable to the Low Variable rectangle at the center left.*

    c. *Drag the Sale Price variable to the Close Variable rectangle at the bottom on the left.*

    d. *Drag the Neighborhood variable to the X-Axis rectangle.*

6. **Click the OK button.**

    The high-low graph shown in Figure 11-11 appears.

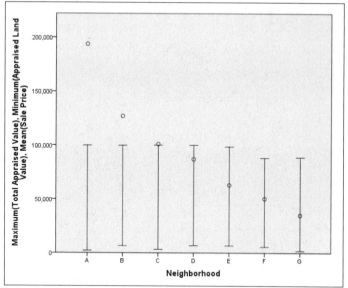

Figure 11-11:
A high-low-
close graph,
displaying a
variable that
curves out
of its high
and low
limits.

Simple range bar graphs

The *simple range bar* graph shows how a variable appears when plotted between high and low values. That is, it displays the relationships among three sets of values. This example and the one before it display the same information, but with a different appearance to the graphics. This graph displays the upper and lower range as a vertical bar.

Do the following to build a simple range bar graph:

1. **Choose File⇨Open⇨Data and open the file named** Home sales [by neighborhood].sav, **which is in the SPSS installation directory.**

2. **Choose Graphs⇨Chart Builder.**

3. **In the Choose From list, select High-Low.**

4. **Drag the second graph diagram (the one with the Simple Range Bar tooltip) to the panel at the top of the window.**

5. **In the Variables list, do the following:**

   a. *Drag the Total Appraised Value variable to the High Variable rectangle at the top on the left.*

   b. *Drag the Appraised Land Value variable to the Low Variable rectangle at the center left.*

   c. *Drag the Sale Price variable to the Close Variable rectangle at the bottom on the left.*

d. Drag the Neighborhood variable to the X-Axis rectangle.

6. Click the OK button.

The high-low range bar graph shown in Figure 11-12 appears.

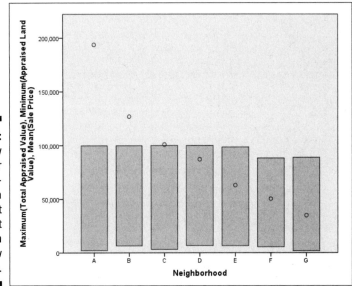

Figure 11-12: A high-low range bar graph, displaying a variable that curves out of its high and low limits.

Clustered range bar graphs

The *clustered range bar* graph displays the relationship among five variables. No other chart can be used to so clearly display so many variables. This example demonstrates the relationships among five employee variables: years of employment, income ranges, years with employer, ages, and the years they've lived at their current addresses.

Do the following to build a clustered range bar graph:

1. Choose File⇨Open⇨Data and open the file named demo.sav, **which is in the SPSS installation directory.**

2. Choose Graphs⇨Chart Builder.

3. In the Choose From list, select High-Low.

4. Drag the third graph diagram (the one with the Clustered Range Bar tooltip) to the panel at the top of the window.

5. In the Variables list, do the following:

a. Drag the category Years With Current Employer[empcat] variable to the Cluster on X rectangle in the upper-right corner.

b. *Drag the Income Category variable to the X-Axis rectangle at bottom.*

c. *Drag the Age In Years variable to the High Variable rectangle at top left.*

d. *Drag the measure Years With Current Employer[employ] variable to the Low Variable rectangle at center left.*

e. *Drag the Years At Current Address variable to the Close Variable rectangle at bottom left.*

6. **Click the OK button.**

The high-low graph shown in Figure 11-13 appears.

Figure 11-13: A clustered range bar graph, displaying relationships among five variables.

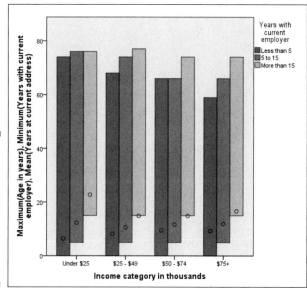

Differenced area graphs

A *differenced area graph* provides a pair of line graphs that emphasize the difference between two variables filling the area between them with a solid color. The two graphs are plotted against the points of a categorical variable. The following steps produce a differenced area graph:

1. **Choose File➪Open➪Data and open the** Home sales [by neighbor-hood].sav **file, which is in the SPSS installation directory.**

2. **Choose Graphs➪Chart Builder.**

3. **In the Choose From list, select High-Low.**

4. **Drag the fourth graph diagram (the one with the Differenced Area tooltip) to the panel at the top of the window.**

5. **In the Variables list, do the following:**

   a. Drag the Neighborhood variable to the X-Axis rectangle.

   b. Drag the Sale Price variable to either of the Y-Axis rectangles.

   c. Drag the Appraised Value of Improvements variable to the other Y-Axis rectangle.

6. **Click the OK button.**

   The differenced area chart shown in Figure 11-14 appears.

Figure 11-14: In a differenced area chart, the filled region emphasizes the difference between two values.

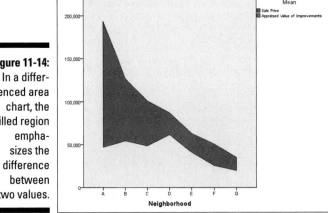

Dual-Axis Graphs

Many of the graphic forms allow you to plot two or more variables on the same chart, but they must always be plotted against the same scale. In the *dual-axis graph*, two variables are plotted — and two different scales are used to plot them. As a result, the values don't require the same ranges (as they do in the other plots); the curves and trends of the two variables can be easily compared, even though they're on different scales.

Dual Y-axes with categorical X-axis

Two variables with different ranges that vary across the same set of categories can be plotted together, as shown in the following example:

1. **Choose File➪Open➪Data and open the** Cars.sav **file, which is in the SPSS installation directory.**

2. **Choose Graphs➪Chart Builder.**

3. **In the Choose From list, select Dual Axes.**

4. **Drag the first diagram (the one with the Dual Y Axes With Categorical X Axis tooltip) to the panel at the top of the window.**

5. **In the Variables list, do the following:**

   a. *Drag the Horsepower variable to the Y-Axis rectangle on the left.*

   b. *Drag the Miles Per Gallon variable to the Y-Axis rectangle on the right, which is now named Count.*

   c. *Drag the Number of Cylinders variable to the X-Axis rectangle.*

6. **Click the OK button.**

   The dual-axis graph shown in Figure 11-15 appears.

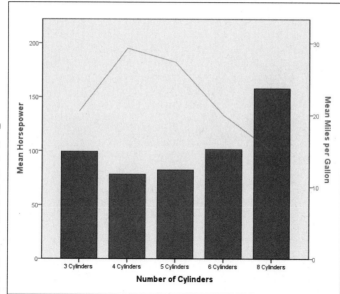

Figure 11-15: A dual-axis graph displaying the curves inscribed by two variables with different ranges.

Dual Y-axes with scale X-axis

If you don't mind a little complexity, two variables with different ranges that vary according to changes in another value laid out along a third scale can be plotted together, as shown in the following example:

1. **Choose File⇨Open⇨Data and open the `Cars.sav` file, which is in the SPSS installation directory.**

2. **Choose Graphs⇨Chart Builder.**

3. **In the Choose From list, select Dual Axes.**

4. Drag the second graph diagram (the one with the Dual Y Axes With Scale X Axis) to the panel at the top of the window.

5. In the Variables list, do the following:

a. Drag the Miles Per Gallon variable to the Y-Axis rectangle on the left.

b. Drag the Engine Displacement variable to the Y-Axis rectangle on the right.

c. Drag the Time to Accelerate 0 to 60 variable to the X-Axis rectangle.

6. Click the OK button.

The dual-axis chart shown in Figure 11-16 appears.

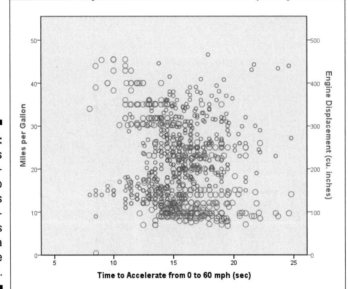

Figure 11-16:
A dual-axis graph displaying two variables with different ranges against a scale variable.

The graph displayed in Figure 11-16 is a combination of two dot-plot formats, with the dots in different colors. Even on a color screen, the two sets of values — each set plotted on a different Y-axis scale — can be confusing. With this type of plot, you must take care that your data makes sense being displayed this way.

Part IV
Analysis

"Unless there's a corrupt cell in our spreadsheet analysis concerning the importance of trunk space, this should be a big seller next year."

In this part . . .

This is the math part. SPSS is so good at it, you can almost hear the numbers crunching. The good news is that you don't have to worry about how to crunch the numbers yourself — all you need to know is how to tell SPSS to crunch the numbers.

Simply select your preferred cruncher, and SPSS does the rest. The output is nicely formatted in a form you can use to impress others. You are the statistical Harry Potter and SPSS is your magic wand.

Chapter 12

Executing an Analysis

● ●

In This Chapter

▶ Generating reports by summarizing data

▶ Displaying summary data in rows and columns

▶ Manipulating the display of pivot tables

● ●

To execute an analysis, you run your numbers through one or more procedures to produce numbers that present a conclusion. The SPSS Viewer displays the output from an analysis in the form of a *pivot table* — so called because you can change them after they're produced, and one dramatic change is to *pivot* the rows so they become columns and the columns so they become rows.

Generating Reports

A report generated in SPSS is created as the result of running an analysis. The analysis can be as simple as specifying how subtotals and totals are to be calculated, or as complex as applying a multipart series of equations.

Computer-generated reports depend on the concept of a *break variable*. If a report will contain subtotals, or some other type of logical internal break, you must define the conditions under which the break will be made. A break usually occurs when a variable changes value. For example, if you're generating a list of employee sick days and want to insert subtotals for male and female, you could use the Gender variable as the break variable: A subtotal could be printed at the end of the 'f' values representing female, and again at the end of the 'm' values representing male.

Processing summaries

When you request that SPSS create a table from your data, you also get another table — the *processing summary*. It appears in SPSS Viewer immediately before the table you requested. Its purpose is to provide you with information about the actions taken by SPSS to create your table. You don't have to request a processing summary to get one.

Figure 12-1 is a simple example of a processing summary. In this example, the values from the Engine Displacement and Horsepower variables in the Cars.sav file were included in the table, which was organized to display information by Miles per Gallon. In a SPSS table, the letter *N* is used as a header to indicate a simple count, or number, of items. The counts tell you how many variables were included. In this example, if all the selected cases had been included in the report, there would have been 406 for each variable. A small number of cases (8 for one variable, 14 for the other) were excluded, so the report included data from 398 cases for one variable and 392 for the other. A case is excluded if the data is missing for a variable. You can see from the table that the percentage of excluded cases is quite small.

Figure 12-1:
A typical
processing
summary
table.

Case Processing Summary

	Cases					
	Included		Excluded		Total	
	N	Percent	N	Percent	N	Percent
Engine Displacement (cu. inches) * Miles per Gallon	398	98.0%	8	2.0%	406	100.0%
Horsepower * Miles per Gallon	392	96.6%	14	3.4%	406	100.0%

Case summaries

You can construct a case summary to organize and summarize the values from one or more variables. Follow these steps:

1. **Choose File⇨Open⇨Data and open the** Cars.sav **file.**

   The file is in the SPSS installation directory.

2. **Choose Analyze⇨Reports⇨Case Summaries.**

   The Summarize Cases dialog box appears.

3. **In the list on the left, do the following:**

   a. *Select Engine Displacement and move it to the Variables panel by clicking the arrow button.*

   b. *Select Horsepower and move it to the Variables panel.*

   c. *Select Miles per Gallon and move it to the Grouping Variable(s) panel.*

   The dialog box should now look like the one in Figure 12-2, with Engine Displacement and Horsepower to be summarized, and the summaries to be grouped by Miles per Gallon. The default, in the lower-left corner of the window, is to limit the summary to the first 100 cases and exclude cases with invalid (missing) values.

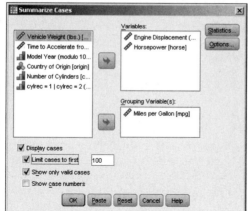

Figure 12-2:
Select the
variables
to include
in the case
summary
table.

4. Click the Statistics button.

The dialog box in Figure 12-3 appears. Here, you can select the statistics you would like to include in the report. The ones available are on the left and the ones selected are on the right.

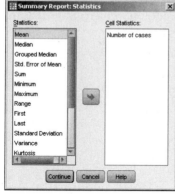

Figure 12-3:
Choose the
ways you
want to
have your
summary
presented.

5. Make certain the only statistic selected is Number of Cases, and then click Continue.

6. Click the Options button.

The dialog box in Figure 12-4 appears. The Title is the text that appears at the top of the table, and the Caption is text that appears at the bottom. In the text you enter for the Title or Caption, you can include \n to split the text to more than one line. You can also use this dialog box to choose whether to have cases with missing values listed in the summary. If you do list them, it is most common to have them appear as periods or asterisks, but you can use any symbol you like.

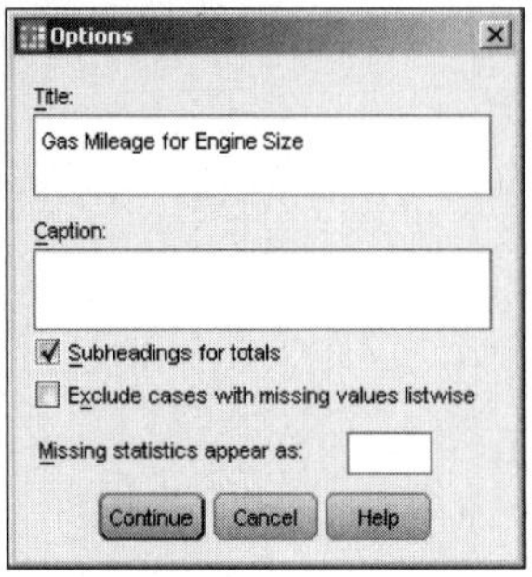

Figure 12-4:
Choose the text to be placed at the top and the bottom of the table.

7. **Replace the default title and click Continue.**

   In this example, replace the default title (Case Summaries) with *Gas Mileage for Engine Size*.

8. **Click OK.**

Figure 12-5 is the top portion of the table produced in this example (the entire table is too large to show conveniently). The table includes mileage data only from the first 100 cases: 2 cars report 10 miles per gallon, 2 report 11 miles per gallon, and 3 report 12 miles per gallon. Each car has its engine displacement and horsepower reported. The small letter *a* appended to the title indicates the presence of a footnote, which states that this report includes only the first 100 cases.

Gas Mileage for Engine Size[a]				Engine Displacement (cu. inches)	Horsepower
Miles per Gallon	9	1		4	93
		Total	N	1	1
	10	1		360	215
		2		307	200
		Total	N	2	2
	11	1		318	210
		2		429	208
		Total	N	2	2
	12	1		383	180
		2		350	160
		3		429	198
		Total	N	3	3
	13	1		400	170
		2		400	175
		3		350	165
		4		350	155
		5		400	190
		6		307	130
		7		302	140

Figure 12-5:
A case summary table.

Summaries in rows

You can produce a report that lists the values of a variable in a column down the left with the values for other variables associated with it in a row to its right. In addition, you can elect to have multiple rows for each break variable by simply selecting the type of statistic.

A row summary table is simple to create but very flexible; it offers lots of options. You'll find a lot of dialog boxes, but the decisions you have to make are easy. After you've run through the process a couple of times to see how it all works, you'll be able to romp through the sequence and produce output without guidance.

The following steps produce a table while giving you a tour of most of the options:

1. **Choose File⇨Open⇨Data and open the** `Cars.sav` **file.**

   The file is in the SPSS installation directory.

2. **Choose Analyze⇨Reports⇨Report Summaries in Rows.**

3. **In the list on the left, do the following:**

   a. *Select Engine Displacement and move it to the Data Column Variables panel by clicking the arrow button.*

   b. *Select Horsepower and move it to the Data Column Variables panel.*

   c. *Select Miles per Gallon and move it to the Break Column Variables panel.*

   The variable names in your dialog box should now look like the ones in Figure 12-6.

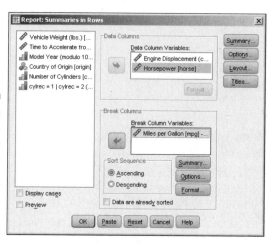

Figure 12-6:
The variables selected to be included in a row summary report.

4. In the Break Columns area, click the Summary button.

This button is enabled only if the Miles per Gallon variable is selected. The dialog box in Figure 12-7 appears.

Figure 12-7:
One row
for each
statistic type
appears
for each
break-vari-
able value.

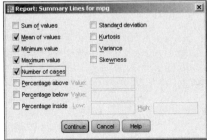

5. Select the Mean of Values, Minimum Value, Maximum Value, and Number of Cases check boxes, and then click Continue.

The report will include a row for each of these types of statistics. When you click the Continue button, the dialog box closes — and the dialog box shown in Figure 12-6 appears again.

6. In the upper-right corner of the dialog box, click the Summary button.

The dialog box in Figure 12-8 appears.

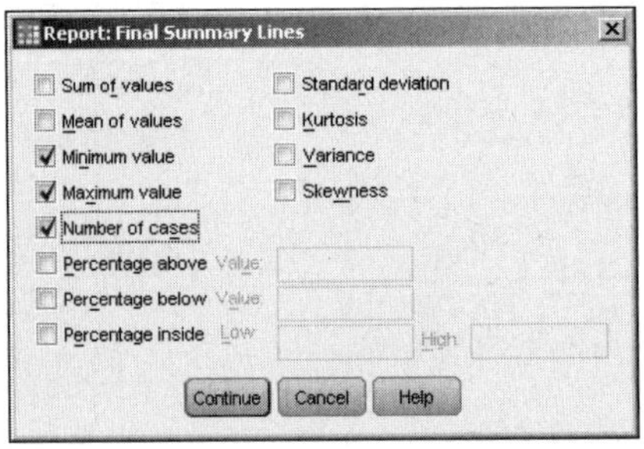

Figure 12-8:
Selection
of the types
of summary
values to
appear at
the bottom
of the table.

7. Select the Minimum Value, Maximum Value, and Number of Cases check boxes, and then click Continue

These values appear as part of the summary at the bottom of the resulting table. (When you click the Continue button, the dialog box closes and the dialog box shown in Figure 12-6 appears again.)

8. In the upper-right corner of the dialog box, click the Options button.

The dialog box in Figure 12-9 appears.

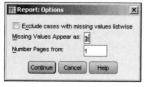

Figure 12-9:
Determine whether —
and how
— to include missing values.

9. **In the Missing Values Appear As text box, type @ (an at sign), and then click Continue.**

   The usual default in this text box is a period. You'll need to replace it with the @ sign. Missing values will be displayed as the character you enter. Alternatively, you could decide to exclude missing values entirely. (When you click the Continue button, the dialog box closes and the dialog box shown in Figure 12-6 appears again.)

10. **In the upper-right corner of the dialog box, click the Titles button.**

    The dialog box in Figure 12-10 appears.

Figure 12-10:
You can define multiple lines of headers and footers.

11. **At the top (in the text box labeled Left), type the text** Miles per Gallon. **Select the Next button above it, and then enter** by Engine Size **in the same text box.**

This specifies that the heading will be two lines in length, and the text on the left will be *Miles per Gallon by Engine Size.* The text on the right of the first line will default to the page number.

12. Click Continue and then click OK.

The output is shown in Figure 12-11. The titles are the text entered in the Titles dialog box (refer to Figure 12-10). The missing value for Miles per Gallon, displayed as @ (as specified in Step 9), occurred in eight cases.

Figure 12-11:
A summary in rows with a custom title and missing data displayed.

```
Miles Per Gallon            Page    1
by Engine Size

Miles            Engine
per           Displacement
Gallon        (cu. inches)  Horsepower
_____        _____  _____


@
Mean                  262        135
Minimum                97         48
Maximum               383        175
N                       8          8

9
Mean                    4         93
Minimum                 4         93
Maximum                 4         93
N                       1          1

10
Mean                  334        208
Minimum               307        200
Maximum               360        215
```

In the output, the break variable is `Miles per Gallon` and appears in the first column. Also in the first column are the names of the types of statistics, and to the right of each one is a row of values for that statistic for each variable chosen — that's why this table is known as summary in rows.

The dialog boxes in this example contain some buttons we didn't use; all of them have to do with formatting details and are self-evident. You can ignore them because the defaults are reasonable, but if you want to make changes to the display, you can do so by clicking the Layout button or either Format button (refer to Figure 12-6). The Format buttons provide options for displaying the currently selected variable.

What the Titles dialog box does, however, may need a bit of explanation (refer to Figure 12-10). The dialog box has two sets of three text boxes. The top set determines the text of each page's title, and the lower set determines the text of each page's footer. You can define as many lines of text for each as you want. The text boxes allow you to define the left, middle, and right of one line. As soon as you enter text for a line, the Next button becomes available and you can click it to move to the text of the next line. The Previous button allows you to back up and make changes.

Summaries in columns

You produce a report in columns by following almost the same procedure used to produce a report in rows. The options are similar, but the form of the report is quite different. You can produce a summary in column format with the following steps:

1. **Choose File⇨Open⇨Data and open the** Home sales [by neighbor-hood].sav **file.**

   The file is in the SPSS installation directory.

2. **Choose Analyze⇨Reports⇨Report Summaries in Columns.**

   The Report Summaries in Columns dialog box appears.

3. **In the list on the left, do the following:**

   a. *Choose Appraised Land Value and move it to the Data Columns panel by clicking the arrow button.*

   It appears with its name and statistic type as landval:sum.

   b. *Select Appraised Value of Improvements and move it to the Data Columns panel.*

   Its name and statistic type appear as improval:sum.

   c. *Select Neighborhood and move it to the Break Columns panel.*

4. **Click the Insert Total button.**

   The word Total (defining a new column) will be added to the bottom of the list in the Data Columns list. Your dialog box should now look like the one in Figure 12-12.

5. **Select** landval:sum **from the Data Columns Variables list and then click Summary.**

   The dialog box in Figure 12-13 appears.

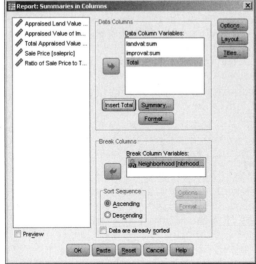

Figure 12-12:
The variables that will appear in the summaries in columns report.

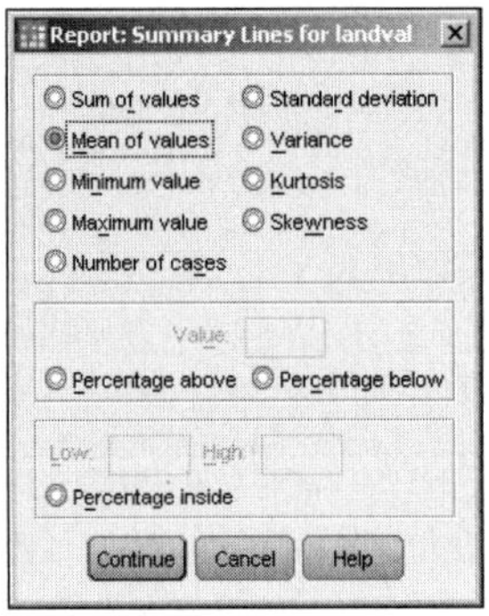

Figure 12-13:
Specify which statistic value will be calculated for a variable.

6. **Select Mean of Values and then click Continue.**

   The first variable in the Data Columns panel is now listed as `landval:mean` to show that the variable is the same as before, but the statistic is now mean instead of sum.

7. **Select Total in the Data Columns Variables panel and then click Summary.**

   The Summary Column dialog box appears.

8. **Select `landval:mean` in the Data Columns panel and click the arrow button to move it to the Summary Column panel (see Figure 12-14). Do the same for `improval:mean`.**

Here you're choosing the variables to be summed to produce the total. You could calculate the total in ways other than a simple sum by selecting another option from the pull-down list, but the default Sum of Columns is right for this example.

Figure 12-14: Select the fields to be summed to create the total.

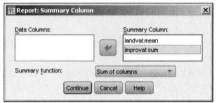

9. **Click Continue and then click OK.**

The table is output and displayed by SPSS Viewer, as shown in Figure 12-15.

			Page 1
Neighborhood	Appraised Land Value Mean	Appraised Value of Improvements Mean	Total
A	44718	47064	91782
B	26671	54302	80974
C	24175	48461	72636
D	18221	61452	79673
E	13729	41048	54777
F	14297	25426	39722
G	8847	19731	28579

Figure 12-15: A summary in columns report with a total column.

For each neighborhood listed in the first column, the report shows the mean land appraisal value, the mean appraisal value of the improvements, and the total of the two means — the total mean appraisal value.

Other options are available for defining the appearance of this report, but the defaults are reasonable and probably should be used unless you have something specific in mind. You can use the Titles button to specify the text of the headers and footers just as you would for summaries in the rows report.

OLAP cubes

A regular table is in two dimensions: height and width. A *cubed table* is a table in three dimensions: height, width, and depth. It's like a deck of cards with a regular two-dimensional table printed on each card. You can flip from one card to another to see any of the tables. Thus it adds the third dimension, depth, and the table becomes cubed.

An *OLAP (Online Analytical Processing) cube* is the output from a process that uses one or more scale variables along with one or more categorical values to divide the report information into layers for the depth. The following steps guide you through the process of producing a three-dimensional table:

1. **Choose File➪Open➪Data and open the** `Employee data.sav` **file.**

   The file is in the SPSS installation directory.

2. **Choose Analyze➪Reports➪OLAP Cubes.**

   The OLAP Cubes dialog box appears, as shown in Figure 12-16.

Figure 12-16: The OLAP Cubes dialog box is where you choose variables for the tables.

3. **In the list on the left, do the following:**

   a. *Select Current Salary and move it to the Summary Variable(s) panel by clicking the arrow button.*

   b. *Select Beginning Salary and move it to the Summary Variable(s) panel.*

   c. *Select Educational Level and move it to the Grouping Variable(s) panel.*

   d. *Select Employment Category and move it to the Grouping Variable(s) panel.*

The results should look like Figure 12-17. These selections will produce a table with several layers: The beginning salary and the current salary will each be shown in separate tables based on educational level and job category. The Statistics button is now available in the dialog box because the variables chosen will make up a valid table.

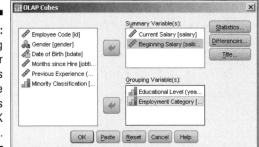

Figure 12-17: Choosing proper variables enables the Statistics and OK buttons.

4. Click the Statistics button.

The OLAP Cubes Statistics dialog box appears, as shown in Figure 12-18. In this dialog box, you decide what calculations you want SPSS to perform.

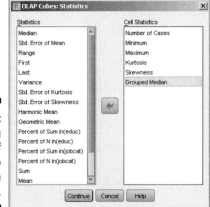

Figure 12-18: Selecting the kinds of statistics to display in the table.

5. Change the list of selected Cell Statistics to include only Number of Cases, Minimum, Maximum, Kurtosis, Skewness, and Grouped Median.

These statistics will be calculated to fill in the table. To select a statistic, highlight its name in the list on the left and click the arrow button to move it to the right. To deselect a statistic, select its name in the list on the right and click the arrow button to move it to the left.

REMEMBER

The order in which the variable names appear in the list determines the order of the values in the table. You determine that order by moving the names into the list in the order in which you want the values to appear. To change the order, you can take the names out and then move them back in the order you want.

6. **Click the Continue button, then the OK button.**

The table in Figure 12-19 appears. This is only the total layer of the multilayered table.

Figure 12-19:
One layer of a multilayered table displaying the statistics for the total.

OLAP Cubes

Educational Level (years):Total
Employment Category:Total

	N	Minimum	Maximum	Kurtosis	Skewness	Grouped Median
Current Salary	474	$15,750	$135,000	5.378	2.125	$28,900.00
Beginning Salary	474	$9,000	$79,980	12.390	2.853	$14,988.68

Double-clicking the OLAP Cubes table selects it and causes the appearance of pull-down lists, as shown in Figure 12-20. One pull-down list appears for each grouping variable. By making selections from the lists, you change the view by changing the table that appears on top.

Figure 12-20:
An OLAP cube stack of tables with two variables determining which table is in view.

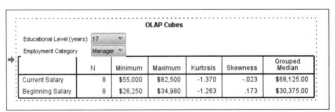

OLAP Cubes

Educational Level (years) 17
Employment Category Manager

	N	Minimum	Maximum	Kurtosis	Skewness	Grouped Median
Current Salary	8	$55,000	$82,500	-1.370	-.023	$68,125.00
Beginning Salary	8	$26,250	$34,980	-1.263	.173	$30,375.00

Modifying Pivot Tables

The tables that appear as output in SPSS Viewer are called *pivot tables* because you can change their appearance in several ways — not the least of which is to pivot the table by swapping the rows with the columns.

To modify a table in SPSS Viewer, the table must first be selected, then activated. You select a table by clicking on its name in the list on the left, or by clicking directly on the table. Selection is signified by the presence of a small, red arrow. To then activate the table, it is necessary to double-click on it. An activated table is designated by being surrounded by a dashed line. Then, clicking the right mouse button on the table (you may actually have to do this a couple of times) brings up a menu with Toolbar as its last selection. Selecting this activates the toolbar shown in Figure 12-21.

Figure 12-21:
Use the
toolbar
to modify
the table's
appearance.

Pivot Controls icon

The controls in this toolbar can be used to change the appearance and layout of the table. The changes are immediate. If you use the controls to specify changes to the font (font size, text position, or whatever) the changes show up immediately in the Viewer.

The most dramatic change is a pivot. To perform a table pivot, click the Pivot Controls symbol (it is third from the left on the Formatting toolbar). The dialog box shown in Figure 12-22 appears.

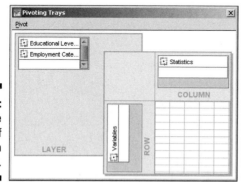

Figure 12-22:
The
positions of
variables in
the table.

As you change the position of variable names in the dialog box, the display of the table in SPSS Viewer changes. You change the positions by dragging the variable names from one place to another.

The box at the upper left initially holds the variable names. The box at the lower right represents the table. You can drag the variable names from the list to the table and drop them on either the rows or the columns. If you don't like the result, you can drag the names back to the list, or to another position in the table. When you drag a variable name, SPSS tells you whether it can be dropped; if it is dropped, the layout of the table shows an immediate change.

Chapter 13

Some Analysis Examples

*T*his chapter describes how to instruct SPSS to dig into your data, execute an analysis, and reach a conclusion. In SPSS, executing an analysis consists of taking in your raw data, performing calculations, and presenting the results in a table or a chart.

This chapter provides examples of the most fundamental types of analysis that SPSS offers. Any menu choices or options that I don't demonstrate here are more advanced forms of the same types of analysis. The more advanced forms often require that you set more options, and sometimes they require the selection of more variables from the dataset, but the process is always fundamentally the same as the examples described in this chapter. The advanced forms employ the same basic algorithms. In general, understanding how the analysis examples in this chapter operate will give you the understanding you need for the more advanced forms.

In the descriptions in this chapter, I assume you're familiar with the fundamental procedures required for constructing tables, which I describe in Chapter 12.

Comparison of Means Analyses

The tests for comparing the mean of one variable to the mean of another are more varied and flexible than you might think. The analysis methods in this section fall into the category of means tests, but they are actually more than that. You'll find that they can produce up to twelve statistics, of which the mean is only one.

Simple means compare

You can generate a simple comparison table by loading the `Employee data.sav` file and choosing Analyze⇨Compare Means⇨Means. The dialog box in Figure 13-1 appears, showing a list of variable names at the left. Select the variables you want to use for calculating the mean — Beginning Salary and Current Salary — and transfer them to the Dependent List panel (by clicking the arrow button). Select the Employment Category variable and move it to the Independent List panel. This is all you have to do to produce output.

Figure 13-1: Choosing the variables that will generate the table.

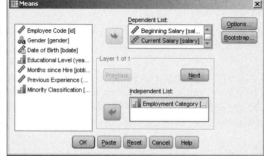

The table produced from this dialog box is shown in Figure 13-2. By default, it includes the means, a count of the number of cases, and the standard deviation. But you are not limited to only these. By clicking the Options button on the dialog box shown in Figure 13-1, you can choose any combination of 21 statistics.

When you run an analysis that produces a table as output, a second table is also produced: the Case Processing Summary table. It provides a quick summary of the number of cases included and omitted. Cases with a missing value for a chosen variable are omitted.

That's not the end of the possible variations. You can include other independent variables in two ways:

- ✔ The table in Figure 13-2 is single-layered, but by clicking the Next button in the dialog box in Figure 13-1, you can add new layers for independent variables.

- ✔ You can add independent variables to the same top layer (or any other layer) and make a larger table to include them.

Report

Employment Category		Beginning Salary	Current Salary
Clerical	Mean	$14,096.05	$27,838.54
	N	363	363
	Std. Deviation	$2,907.474	$7,567.995
Custodial	Mean	$15,077.78	$30,938.89
	N	27	27
	Std. Deviation	$1,341.235	$2,114.616
Manager	Mean	$30,257.86	$63,977.80
	N	84	84
	Std. Deviation	$9,980.979	$18,244.776
Total	Mean	$17,016.09	$34,419.57
	N	474	474
	Std. Deviation	$7,870.638	$17,075.661

Figure 13-2:
Comparison of means and standard deviation according to employment category.

One-sample T test

The *one-sample T test* compares an expected value with a mean derived from the values of a single variable. To run such a test, you choose the variable you want to average and the value you expect. The report shows you the accuracy of your expectations.

For an example of the T test, open the `Employee data.sav` file. Choose Analyze⇨Compare Means⇨One Sample T Test and the dialog box in Figure 13-3 appears. As shown in Figure 13-3, I selected the Educational Level variable and the number 12. The mean of the variable will be compared against the constant value 12.

Figure 13-3:
Select a variable and the value you think its mean should have.

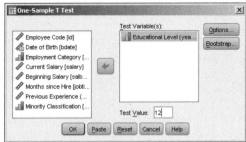

The resulting table is shown in Figure 13-4. At the top of the table is the value that's the basis of all comparisons — the average number of years of education of all employees was compared to 12. The first column, labeled with the letter *t*, is the mean value derived from the data. The second column, the one labeled *df*, is the degrees of freedom. The Mean Difference column is the average of the magnitude of the differences between the values and the expected value. The Confidence Interval values show how wide the range is around the value of 12 to include 95 percent of all values.

Figure 13-4:
T test comparing 12 with the mean of years of education.

One-Sample Test						
	Test Value = 12					
					95% Confidence Interval of the Difference	
	t	df	Sig. (2-tailed)	Mean Difference	Lower	Upper
Educational Level (years)	11.257	473	.000	1.492	1.23	1.75

Independent-samples T test

The *independent-samples T test* compares the means of two sets of values from one variable. To run an example of the test, load the `Employee data. sav` file. Choose Analyze⇨Compare Means⇨Independent-Samples T Test, and the dialog box in Figure 13-5 appears.

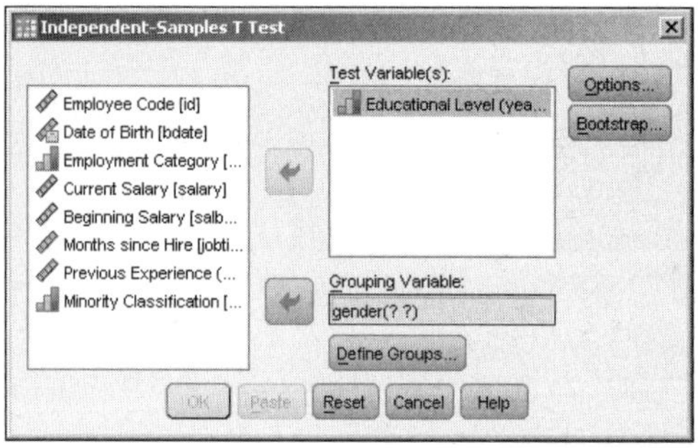

Figure 13-5:
Test to compare the means of two variables.

Move the Educational Level variable to the Test Variable(s) panel. This variable supplies the values for the means you want to test. Move the Gender variable to the Grouping Variable panel; this variable is used to select the two groups. The variable could have multiple values defined for it, but you need to choose only two. Click the Define Groups button to specify the two values — in this example, the only values available are m and f. Entering these two values causes them to appear in place of the question marks following the name of the variable. Click the OK button, and SPSS produces the pair of tables shown in Figure 13-6.

The Independent Samples Test table displays the two means, the standard deviation and standard error for the two means. The table provides further information about the mean in two rows of numbers, but you need to know some things about reading the table. The table has one row for equal variances and one for unequal variances. You decide which to use this way:

- ✔ If the significance of the Levene test — the number in the second column — is high (greater than 0.05 or so), the values in the first row are applicable.

- ✔ If the significance of the Levene test is low, the numbers in the second row are more applicable.

- ✔ If the significance of the T test — that is, the two-tailed significance — is low, it indicates a significant difference in the two means.

- ✔ If none of the numbers of the 95 percent confidence interval are 0, it indicates that the difference between the means is significant.

Figure 13-6:
The pair of tables produced from the independent-samples T test.

Group Statistics

	Gender	N	Mean	Std. Deviation	Std. Error Mean
Educational Level (years)	Male	258	14.43	2.979	.185
	Female	216	12.37	2.319	.158

Independent Samples Test

		Levene's Test for Equality of Variances		t-test for Equality of Means						95% Confidence Interval of the Difference	
		F	Sig.	t	df	Sig. (2-tailed)	Mean Difference	Std. Error Difference	Lower	Upper
Educational Level (years)	Equal variances assumed	17.884	.000	8.276	472	.000	2.060	.249	1.571	2.549
	Equal variances not assumed			8.458	469.595	.000	2.060	.244	1.581	2.538

Paired-samples T test

The *paired-samples T test* is a comparison test specially designed to compare values from the same group at different times. The values could be gathered before and after an event, or simply before and after a passage of time.

To run the test, choose Analyze⇨Compare Means⇨Paired-Samples T Test. Select Current Salary from the list on the left, click the arrow button, and then it shows up on the right. Do this again with Beginning Salary, and you get a pair of variables, as shown in Figure 13-7. If you wish, you can select more than one pair and run more than one test at a time. You can use the double-headed arrow button to swap variable 1 with variable 2 in any pair. That's all there is to it, unless you want to use the Options button to change the 95 percent confidence level to another percentage. Click the OK button to produce the paired-samples T test table.

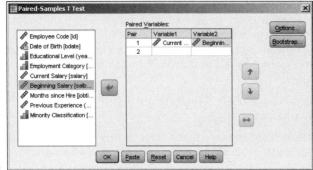

Figure 13-7:
Selecting
two vari-
ables
causes
them to
appear on
one line in
the panel.

One-way ANOVA

ANOVA is an acronym for ANalysis Of VAriance. A one-way ANOVA is the
analysis of the variance of values (of a dependent variable) by comparing
them against another set of values (the independent variable). It is a test of
the hypothesis that the mean of the tested variable is equal to that of the
factor.

The output table from running this test is a small one. To see an example
of its output, load the Road construction bids.sav file. Then choose
Analyze⇨Compare Means⇨One-Way ANOVA. In the dialog box shown in
Figure 13-8, I'm testing the hypothesis that the mean of the contractor's construction costs matches that of the department of transportation's engineer-
ing cost estimates. The result is the table shown in Figure 13-9.

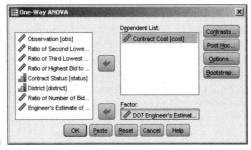

Figure 13-8:
One variable
is chosen
to be tested
and another
is chosen as
the factor to
test against.

Figure 13-9:
The analysis
of the vari-
ance of one
variable as
compared
to that of
another.

ANOVA

Contract Cost

	Sum of Squares	df	Mean Square	F	Sig.
Between Groups	...	233	3806325.945	3704.798	.013
Within Groups	1027.404	1	1027.404		
Total	...	234			

Linear Model Analyses

Many statistical values result from comparing actual results against expected results — or, in statistics-speak, the comparison of dependent variables against independent variables. Straight lines are easier to compare than curves and often produce a result that's easier to understand. This section is about curveless analysis.

One variable

You can compare one dependent variable against more than one independent variable. For example, suppose a plastic manufacturer wants to increase the resistance to tearing of his product, so he varies the extrusion rate and additives to do so. To see how the results of the study can be calculated, open the `Plastic.sav` file. Then choose Analyze➪General Linear Model➪Univariate.

Select the Tear Resistance variable to be the one dependent variable, and the Additive Amount and Extrusion variables to be the fixed variables, as shown in Figure 13-10.

The table in Figure 13-11 is produced, displaying the resulting values of Tear Resistance, depending on Extrusion and Additive Amount, both individually and together.

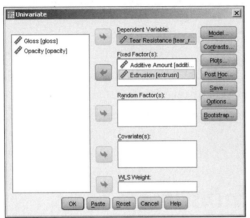

Figure 13-10:
One
dependent
variable
is tested
against
more than
one inde-
pendent
variable.

Figure 13-11:
The Tear
Resistance
variable
is tested
against the
effect of two
factors.

Tests of Between-Subjects Effects

Dependent Variable:Tear Resistance

Source	Type III Sum of Squares	df	Mean Square	F	Sig.
Corrected Model	2.501[a]	3	.834	7.563	.002
Intercept	920.724	1	920.724	8351.243	.000
additive	.760	1	.760	6.898	.018
extrusn	1.740	1	1.740	15.787	.001
additive * extrusn	.000	1	.000	.005	.947
Error	1.764	16	.110		
Total	924.990	20			
Corrected Total	4.265	19			

a.R Squared = .586 (Adjusted R Squared = .509)

More than one variable

It is also possible to measure more than one dependent variable against more than one independent variable. Using the same data as in the single-value test of the preceding section, choose Analyze⇨General Linear Model⇨Multivariate. The Gloss, Tear Resistance, and Opacity dependent variables will be tested against the Additive Amount and Extrusion variables, as shown in Figure 13-12.

Click the OK button, and the table in Figure 13-13 is produced. You may notice that this table is the same basic form as the single-value table in the preceding section, except the Dependent Variable column now has three entries for each entry in the Source column.

Figure 13-12:
Three
dependent
variables
are tested
against two
independent
variables.

Tests of Between-Subjects Effects

Source	Dependent Variable	Type III Sum of Squares	df	Mean Square	F	Sig.
Corrected Model	Gloss	2.457[a]	3	.819	4.987	.012
	Tear Resistance	2.501[b]	3	.834	7.563	.002
	Opacity	9.282[c]	3	3.094	.762	.531
Intercept	Gloss	1735.385	1	1735.385	10565.507	.000
	Tear Resistance	920.724	1	920.724	8351.243	.000
	Opacity	309.684	1	309.684	76.319	.000
additive	Gloss	.612	1	.612	3.729	.071
	Tear Resistance	.760	1	.760	6.898	.018
	Opacity	4.900	1	4.900	1.208	.288
extrusn	Gloss	1.301	1	1.301	7.918	.012
	Tear Resistance	1.740	1	1.740	15.787	.001
	Opacity	.420	1	.420	.104	.752
additive * extrusn	Gloss	.544	1	.544	3.315	.087
	Tear Resistance	.000	1	.000	.005	.947
	Opacity	3.960	1	3.960	.976	.338
Error	Gloss	2.628	16	.164		
	Tear Resistance	1.764	16	.110		
	Opacity	64.924	16	4.058		
Total	Gloss	1740.470	20			
	Tear Resistance	924.990	20			
	Opacity	383.890	20			
Corrected Total	Gloss	5.085	19			
	Tear Resistance	4.265	19			
	Opacity	74.206	19			

a. R Squared = .483 (Adjusted R Squared = .386)
b. R Squared = .586 (Adjusted R Squared = .509)
c. R Squared = .125 (Adjusted R Squared = -.039)

Figure 13-13:
Tear
Resistance,
Gloss, and
Opacity all
being tested
against the
effect of two
factors.

Correlation Analyses

The group of tests in this section determines the similarity or difference in the way two variables change in value from one case (row) to another through the data.

Bivariate

To run a simple *bivariate* (two-variable) correlation, load data that has two variables to be compared and choose Analyze⇨Correlate⇨Bivariate. In Figure 13-14, I'm using `Employee data.sav` to perform a test to that determines whether there's a correlation between an employee's starting salary and current salary.

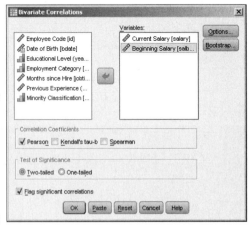

Figure 13-14: Select variables to be compared by moving them to the right.

You can choose up to three kinds of correlations. The most common form is the Pearson correlation, which is the default. If you want, you can click the Options button and decide what is to be done about missing values and to tell SPSS whether you want to calculate the standard deviations. The result of the selections in Figure 13-14 is shown in Figure 13-15.

Figure 13-15: Pearson correlation showing a highly significant correlation.

Correlations

		Current Salary	Beginning Salary
Current Salary	Pearson Correlation	1	.880**
	Sig. (2-tailed)		.000
	N	474	474
Beginning Salary	Pearson Correlation	.880**	1
	Sig. (2-tailed)	.000	
	N	474	474

**.Correlation is significant at the 0.01 level (2-tailed).

Correlation figures vary from –1 to +1, and the larger the value, the stronger the correlation. In Figure 13-15, you can see that the variables have a correlation of 1 with themselves and .880 with one another, which is a significant correlation.

Partial correlation

Outside factors can affect a correlation. You can include such factors in the calculations by using a procedure known as a *partial correlation*. For example, in the previous example, I found that the current salary of each employee correlated with the starting salary, but I did not take into account the length of employment. In this example, I will. Begin by choosing Analyze➪Correlate➪Partial.

Select the Current Salary and Beginning Salary as the Variables, along with the Months Since Hire as the Controlling factor that should have an effect on the correlation. The resulting dialog box should look like the one in Figure 13-16.

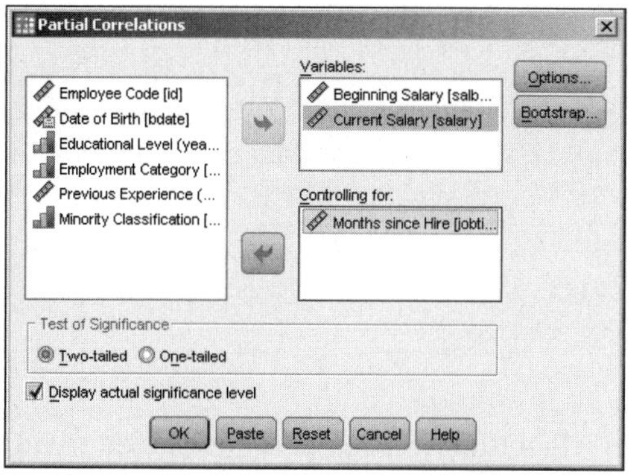

Figure 13-16:
Select the variables to correlate and the variable to control the correlation.

The result is an even higher level of correlation than in the previous example, as shown in Figure 13-17.

Figure 13-17:
The correlation of starting with the current salary and taking the length of employment into account.

Correlations

Control Variables			Beginning Salary	Current Salary
Months since Hire	Beginning Salary	Correlation	1.000	.885
		Significance (2-tailed)	.	.000
		df	0	471
	Current Salary	Correlation	.885	1.000
		Significance (2-tailed)	.000	.
		df	471	0

Regression Analyses

Regression analysis is about predicting the future (the unknown) based on data collected from the past (the known). Such an analysis determines a mathematical equation that can be used to figure out what will happen, within a certain range of probability:

✔ The analysis is performed based on a single dependent variable.

✔ The process takes into consideration the effect one or more dependent variables have on the dependent variable.

✔ It takes into account which independent variables have more effect than others.

Performing regression analysis is the process of looking for predictors and determining how well they predict a future outcome.

When only one independent variable is taken into account, the procedure is called a *simple regression*. If you use more than one independent variable, it's called a *multiple regression*. All dialog boxes in SPSS provide for multiple regression.

Linear

Linear regression is used when the projections are expected to be in a straight line with actual values. The following is an example of a linear multiple regression:

1. **Choose File⇨Open⇨Data and open the** demo.sav **file.**

   The file is in the SPSS installation directory.

2. **Choose Analyze⇨Regression⇨Linear.**

   The Linear Regression dialog box appears.

3. **Select Household Income In Thousands and move it to the Dependent panel.**

   This is the variable for which we want to set up a prediction equation.

4. **Select Level of Education and move it to the Independent(s) panel. And select Age in Years and move it to the Independent(s) panel.**

   The screen should look like Figure 13-18. Here the assumption is that two variables have a linear effect on Household Income in Thousands.

5. **Click OK.**

   The table in Figure 13-19 is produced.

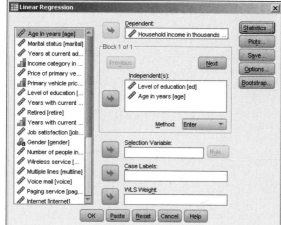

Figure 13-18: Select the variable to be predicted and the independent variables that affect it.

Figure 13-19: The table containing coefficients for making revenue predictions.

Coefficients[a]

Model		Unstandardized Coefficients B	Unstandardized Coefficients Std. Error	Standardized Coefficients Beta	t	Sig.
1	(Constant)	-49.553	4.052		-12.228	.000
	Level of education	9.252	.771	.141	11.998	.000
	Age in years	2.261	.075	.353	30.058	.000

a. Dependent Variable: Household income in thousands

You will also find other tables included as part of the output, but they all have to do with how the values of this table are produced. This table defines the equation for you in its first column. Assuming that the assumption of the linear relationship is correct, income can be predicted with the following:

```
Household Income = (9.252)(Level of Education) + (2.261)
          (Age in Years) - 49.553
```

Curve estimation

This is *non*-linear regression. If you have a collection of data points, it's possible to create a curve that passes through (or very near) those points. The equation of the curve can then be used to estimate the values of points you don't have yet. This can be done by *interpolation* (drawing a curve that connects the existing points) or *extrapolation* (extending the curve beyond the existing points). The graphic presentation of values isn't as numerically accurate as a table of numbers, but it has some advantages — not the least of which is that you can quickly spot patterns and trends. Predictions are only estimations, no matter how sophisticated — so presenting a prediction as a graph is as good as doing so with numbers, given the inherent inexactness of graphic representation of values.

In the following, I fit a curve to a group of data points for the purpose of demonstrating the probable horsepower of an engine, depending on its cubic inches of displacement:

1. **Choose File⟹Open⟹Data and open the `Cars.sav` file.**

   The file is in the SPSS installation directory.

2. **Choose Analyze⟹Regression⟹Curve Estimation.**

   The Curve Estimation dialog box appears.

3. **Select Horsepower as the variable to have its value predicted by moving it to the Dependent(s) panel.**

   You could choose more than one dependent variable — in which case, the output would appear as more than one chart. Each dependent variable has its own graph.

4. **Select Engine Displacement and move it to the Independent panel.**

5. **Select Linear, Quadratic, and Cubic as the types of curves to be generated.**

   The screen should look like Figure 13-20.

6. **Click OK.**

   SPSS generates some tables to describe the processing used to reach its conclusion. The graph shown in Figure 13-21 contains the three requested curves.

In Figure 13-21, each dot represents the relationship of actual engine displacement to measured horsepower. The predicted values of horsepower according to displacement are represented in three ways:

 ✔ The *linear interpretation* is the best fit of a straight line to the dots.

 ✔ The *quadratic line* is the best fit of a line that curves in one direction.

> ✔ The *cubic line* reverses the direction of its curve in an attempt to fit as closely as possible.

None of the curves fit the data points exactly, but they do give you the best possible prediction of the results.

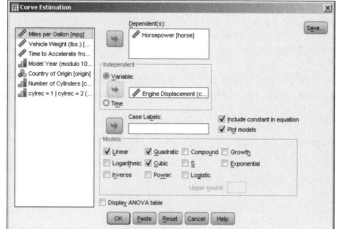

Figure 13-20:
Select the variables involved in curve fitting and the types of curves.

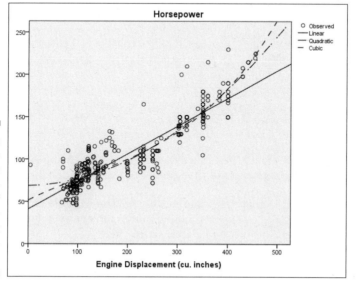

Figure 13-21:
Curves generated from the relationships of engine displacement to horsepower.

Log Linear Analyses

Log linear is based on the assumption that a linear relationship exists between the independent variables and the logarithm of the dependent variable.

The example in this section summarizes the starting salaries of employees, organizing their educational level and employment category. To generate this table, open the `Employee data.sav` file. Then choose Analyze⇨Loglinear⇨General.

Move the Educational Level and Employment Category variables to the Factor(s) panel, making them the two variables used to divvy up the results. Move the Beginning Salary variable to the Contrast Variable(s) panel, making it the variable containing the data to be divvied up. Your screen should look like Figure 13-22.

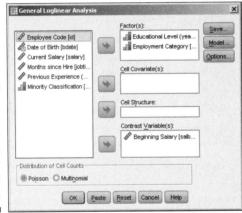

Figure 13-22:
Select the variables to take part in a log linear general analysis.

Click the OK button and the table shown in Figure 13-23 appears. You can see that the salaries increase with the number of years of education.

Executing the same analysis — but leaving out the variable for the Employment Category — we get a table that organizes salaries only by the level of education, as shown in Figure 13-24.

Coefficients[b,c]

Educational Level (years)	Employment Category	Beginning Salary[a]
8	Clerical	$12,304
	Custodial	$15,404
	Manager	missing
12	Clerical	$13,013
	Custodial	$14,758
	Manager	$33,750
14	Clerical	$15,625
	Custodial	missing
	Manager	missing
15	Clerical	$15,262
	Custodial	$15,000
	Manager	$25,433
16	Clerical	$17,564
	Custodial	missing
	Manager	$25,613
17	Clerical	$16,750
	Custodial	missing
	Manager	$30,713
18	Clerical	$27,615
	Custodial	missing
	Manager	$33,561
19	Clerical	$19,500
	Custodial	missing
	Manager	$35,351
20	Clerical	missing
	Custodial	missing
	Manager	$36,240
21	Clerical	missing
	Custodial	missing
	Manager	$37,500

a.Sum of the coefficients is not zero. The generalized log-odds ratio is not computed.

b.Model: Poisson

c.Design: Constant + educ + jobcat + educ * jobcat

Figure 13-23: Starting salaries divided by sex and further divided by the type of degree.

Coefficients[b,c]

Educational Level (years)	Beginning Salary[a]
8	$13,064
12	$13,242
14	$15,625
15	$15,611
16	$22,338
17	$26,905
18	$32,240
19	$34,764
20	$36,240
21	$37,500

a.Sum of the coefficients is not zero. The generalized log-odds ratio is not computed.

b.Model: Poisson

c.Design: Constant + educ

Figure 13-24: Starting salaries grouped by sex and no other factors.

Part V
Programming SPSS with Command Syntax

The 5th Wave   By Rich Tennant

"The top line represents our revenue, the middle line is our inventory, and the bottom line shows the rate of my hair loss over the same period."

In this part . . .

Deep down inside IBM SPSS Statistics, where its heart beats, everything happens because of statements written in the Command Syntax language. With Syntax (as it's known to its friends), you can skip the menus and dialog boxes and issue commands directly to the internals of SPSS. It may sound a bit spooky at first, but it isn't as hard as it sounds. In fact, SPSS will help you write your own Command Syntax statements.

Chapter 14

The Command Syntax Language

*E*verything that happens in SPSS is the result of executing a Command Syntax script. Whenever you use the menu to specify a set of options and then click an OK button that instructs SPSS to perform some feat, a Command Syntax script is generated and executed. You've seen examples of the Command Syntax language — Syntax for short — appearing at the top of the SPSS Viewer every time a command runs. This chapter describes some language fundamentals, and the next one explains how you can write your own programs. These two chapters give you a look at the operation of the Syntax language.

Commands

A single Syntax language instruction can be very simple, or it can be complex enough to serve as an entire program. A single instruction consists of a command followed by arguments to modify or expand the actions of the command. For example, the following Syntax command generates a report:

```
REPORT /FORMAT=LIST /VARIABLES=MPG.
```

The first thing you probably noticed is that the command is written in all uppercase. That's tradition — not a requirement. You can write in lowercase (or even mixed case) if you want. Notice also that the end of the list of arguments is terminated by a single period. The terminator must be there or SPSS will complain.

Now, about those forward slashes and equal signs: Sometimes you need them, and sometimes they're optional. Always use them and you won't have any trouble. The presence of slashes and equal signs reduces ambiguity for both you and SPSS. Also, commands can be abbreviated as long as you have at least three letters that uniquely identify the command. But I can't think of a single reason to abbreviate anything. Figuring out how to abbreviate a command is more work than just typing it, and abbreviation makes the program harder to figure out later.

The command in this example is REPORT, which causes text to be written to SPSS Viewer. In fact, all output produced by running Syntax programs goes to SPSS Viewer. The FORMAT specification tells REPORT to make a list of the values. The VARIABLES specification tells REPORT which variables are to be included in the list.

Commands can begin anywhere on a line and continue for as many lines as necessary. That's why SPSS is so persnickety about that terminator (the period) — it's the only way it has of detecting the end of a command. The maximum length of a single line is 80 characters.

Keywords

All the commands in Syntax are keywords in the language. A *keyword* is a word already known to the language and has a predefined action. The variable names you define are not keywords, but SPSS can tell which is which by the way you use them. That is, you can name one of your variables the same name as one of the keywords, and SPSS can tell what you mean by how you use the word. Usually.

The names of commands, subcommands, and functions are all keywords — and there are lots of them — but they are not reserved and you can use them freely. For example, you could have variables named format and report, and you could use the following Syntax command to display a list of their values:

```
REPORT /FORMAT=LIST /VARIABLES=REPORT FORMAT.
```

Don't try to name variables AND, OR, or NOT. These logical operators are keywords in the Syntax language and are *also* reserved words. If you try to use a reserved word as a variable name, SPSS will catch it and tell you that you can't do it. Relational operators are used in the Syntax language to compare values and are also reserved words. The relational operators are EQ, NE, LT, GT, LE, and GE. Some other reserved words are ALL, BY, TO, and WITH. (These operators are discussed in more detail later in this chapter.)

Variables and Constants

Most of the values used in Syntax come from the variables in the dataset you currently have loaded and displayed in SPSS. You simply use your variable names in your program, and SPSS knows where to go and get the values. Some other variables are already defined, and you can use them anywhere in a program. Predefined variables, which are called *system variables,* all begin with a dollar sign ($) and already contain values. The system variables are listed in Table 14-1.

Table 14-1	System Variables
Variable Name	*Description*
$CASENUM	The current case number. It's the count of cases from the beginning case to the current one.
$DATE	The current date in international date format with a two-digit year.
$DATE11	The current date in international date format with a four-digit year.
$JDATE	The count of the number of days since October 14, 1582 (the first day of the Gregorian calendar).
$LENGTH	The current page length.
$SYSMIS	The system missing value. This prints as a period or whatever is defined as the decimal point.
$TIME	The number of seconds since midnight October 14, 1582 (the first day of the Gregorian calendar).
$WIDTH	The current page width.

You can create variables of your own to use as work areas to hold values while your program is running. These are called *scratch variables.* To create a scratch variable, use the # character at the beginning of the name. For example, the following command displays the number 34:

```
COMPUTE #FRED = 34.
PRINT / #FRED.
EXECUTE.
```

The COMPUTE statement executes once. It looks for the name #FRED, finds that it doesn't exist, so it creates it. Then it puts the value 34 into it. The PRINT command executes one time for each case (row) in the currently loaded dataset, so it prints a line for each case. Each time it executes, it finds that the scratch variable #FRED already exists, so it uses its value. For example, if the dataset contains 87 cases, the number 34 would be printed 87 times. If you were to include a variable name from the dataset with the PRINT statement, each of the values of that variable would be printed. An EXECUTE statement is necessary following some commands — it's explained in detail later.

When a Syntax program executes, it is associated with the currently loaded dataset and uses its variable names and values.

Data Declaration

You can define variables and their values in your program. To do so, you create a DATA LIST, which defines the variable names, and follow it with the list of values between BEGIN DATA and END DATA commands. The following example creates three variables (ID, SEX, and AGE) and fills them with four instances of data:

```
DATA LIST / ID 1-3 SEX 5 (A) AGE 7-8.
BEGIN DATA.
001 m 28
002 f 29
003 f 41
004 m 32
END DATA.
PRINT / ID SEX AGE.
EXECUTE.
```

The DATA LIST command defines the variables. The first variable is ID. Its values are found in the input stream in columns 1 through 3; therefore it's defined as being three digits long. It has no type definition, so it defaults to numeric. The second variable is named SEX. It is one character long, and its values are in column 5 of the input. Its type is declared as alpha (A), so it's declared as a one-character string. The third variable, AGE, is two digits long, is a numeric value, and has its values in columns 7 and 8 of the input.

The BEGIN DATA command comes immediately after the DATA LIST command and marks the beginning of the lines of data — each line is a case. If you've ever wondered what it was like to place data on punched cards, this is it. The fundamental design of SPSS is that old. This form of data entry still works, but this is the old way of getting data into a program. When this list of commands is executed, a normal SPSS window appears, showing a dataset with the variable names and values.

You can do all your processing this way, if you prefer. But you don't have to do it by column numbers. You can enter the data in a comma-separated list, as follows:

```
DATA LIST LIST (',') / ID SEX AGE.
BEGIN DATA.
1,1,28
2,2,29
3,2,41
4,1,32
END DATA.
PRINT / ID SEX AGE.
EXECUTE.
```

END DATA must begin in the first column of a command line. It's the only command in Syntax that has this requirement.

Comments

You can insert descriptive text, called a *comment*, into your program. This text doesn't do anything except help clarify how the program works when you read (or somebody else reads) your code. You start a comment the same way you start any other command: on its own line, using the keyword COMMENT or an asterisk. The comment is terminated by a period. Here's an example:

```
COMMENT This is a comment and will not be executed.
```

An asterisk can be used with the same result:

```
* This is a comment placed here for the purpose of
describing what is going on, and it continues until
it is terminated.
```

You can also put comments on the same line as a command by surrounding them with /* and */. A comment like this can be inserted anywhere inside the command where you'd normally put a blank. For example, you could put a comment at the end of a command line like this:

```
REPORT /FORMAT=LIST /VARIABLES=SALARY /* The comment */.
```

It is important to note that the command is terminated with a period, but the period comes after the comment because the comment is part of the statement.

The Execution of Commands

Commands are executed one at a time, starting from the top of the program. The order is important. In particular, if a variable has not been created yet, you can't use it. For the most part, the order is intuitive; you don't have to think much about what exists and what doesn't.

Some statements don't execute right away. Instead, they're stored for later execution. This is normally of no consequence because the statements will be executed when their result is needed. But you should be aware this is going on because it can cause surprises in some circumstances. For example, the PRINT command has a delayed execution:

```
PRINT / ALL.
```

This is a command to print the complete list of values for every case in your dataset. It will print all the values, or by naming variables it can be instructed to print values of only the ones you choose. However, the PRINT command doesn't do it right away. It stores the instruction for later. When your program comes to a command that executes immediately, the stored commands are executed first. That works fine as long as there's another statement to be executed, but if the PRINT statement is the last one in your program, nothing happens. That is, nothing happens until you run another program, and then the stored statement becomes the first one executed.

But there is an easy fix. All you need to do is end your program this way:

```
PRINT / ALL.
EXECUTE.
```

All the EXECUTE command does is execute any statements that have been stored for future execution. For the PRINT command there is another option. The LIST command does the same thing the PRINT command does, but it executes immediately instead of waiting until the next command:

```
LIST / ALL.
```

This execution delay may seem odd at first, but there's a reason for it: Many commands execute once for each case in your data. For example, if you have a series of three statements and you'd like a combination of the three executed once for each case, you need only enter the commands in your program in series. The group of commands will be stored and then executed, once for each case.

Flow Control and Conditional Execution

Unless you specify otherwise, a program starts at the top and executes one statement at a time through your program until it reaches the bottom, where it stops. But you can change that. Situations come up where you need to execute a few statements repeatedly, or maybe you want to conditionally skip one or more statements. In either case you want program execution to jump from one place to another under your control.

IF

You use the IF command when you have a single statement you want to execute only if conditions are right. For example:

```
IF (AGE > 20) GROUP=2.
```

This statement asks the simple question of whether AGE is greater than 20. If so, the value of GROUP is set to 2. We could have used the GT keyword in place of the > symbol. Table 14-2 lists the relational operators you can use to compare numbers.

Table 14-2		Relational Operators
Symbol	*Alpha*	*Definition*
=	EQ	Is equal to
<	LT	Is less than
>	GT	Is greater than
<>	NE	Is not equal to
<=	LE	Is less than or equal to
>=	GE	Is greater than or equal to

You can also combine the relational expressions with logical operators to ask longer and more complex questions. Here's an example:

```
IF (AGE > 20 AND SEX = 1) GROUP=2.
```

This statement asks whether AGE is greater than 20 and SEX is equal to 1. If so, GROUP is set to 2. The logical operators are listed in Table 14-3.

Table 14-3		Logical Operators
Symbol	**Alpha**	**Definition**
&	AND	Both relational operators must be true.
|	OR	Either relational operator can be true.
~	NOT	Reverses the result of a relational operator.

You should use parentheses to organize expressions so there is no ambiguity about what is being compared. When you construct a complicated conditional expression, it's easy to lose track of your original line of scrimmage.

You have to write your expressions so the computer knows what you're talking about. Spell them out. For example, IF (A LT B OR GT 5) is not valid. It can be written IF ((A LT B) OR (A GT 5)), which is a longer form but has a clearer meaning.

You can compare strings to strings and numbers to numbers, but you can't compare strings to numbers.

DO IF

The DO IF statement works the same way as the IF statement, but with DO IF you can execute several statements instead of just one. Because you can enter several statements before the terminating END IF, the END IF is required to tell SPSS when the DO IF is over. Following is an example with two statements:

```
DO IF (AGE < 5).
COMPUTE YOUNG = 1.
COMPUTE SCHOOL = 0.
END IF.
```

In addition to having the option of including a number of statements at once, you can use DO IF to test several conditions in a series — and execute only the statements of the first true condition(s) by using ELSE IF:

```
DO IF (AGE < 5).
COMPUTE YOUNG = 1.
ELSE IF (AGE < 9).
COMPUTE YOUNG = 2.
ELSE IF (AGE < 12).
COMPUTE YOUNG = 3.
END IF.
```

SELECT IF

The SELECT IF statement is not really flow control, but it works that way. You can use it to remove specific cases, and, as a result, include only the cases you want in your analysis. For example, the following sequence of commands prints only the salary values greater than 40,000:

```
SELECT IF (SALARY > 40000).
PRINT / SALARY.
EXECUTE.
```

Any of the logical operators and relational operators that can be used in other IF statements can be used in SELECT IF statements.

DO REPEAT

If you want to perform a transformation on every value of a variable in a dataset, the easiest way is to use DO REPEAT. For example, here's code that increases the salary in every case by 10 percent:

```
DO REPEAT S=SALARY.
COMPUTE S=S + (S * 0.1).
END REPEAT.
PRINT / SALARY.
EXECUTE.
```

On the DO REPEAT command, the name S is assigned as a stand-in for the values of the SALARY variable. The commands between DO REPEAT and END REPEAT are executed once for each instance of SALARY — that is, once per case. Because S is the stand-in for each value of SALARY, any change you make to S is a change to one of the values of SALARY. At the end of this loop, every value of SALARY is printed.

Several lines with commands can be included between DO REPEAT and END REPEAT. Also, you can use several types of commands inside the loop, including IF, DO IF, and LOOP.

LOOP

With the LOOP command, you execute the same block of one or more statements repeatedly for a counted number of times. Following is a simple loop:

```
LOOP #LC = 1 TO 5.
COMPUTE #COUNT = #COUNT + 1.
END LOOP.
PRINT / #COUNT.
EXECUTE.
```

The first statement is a LOOP command and the scratch variable #LC is defined as a loop counter that runs from 1 to 5. The content of the loop defines another scratch variable, #COUNT, and adds 1 to it. Whenever a new scratch variable is defined, its original value is 0. Each time through the loop, 1 is added to #COUNT, so the value at the end of the loop — the value displayed by the PRINT statement — is 5.

But it doesn't stop there. This entire program is executed once for each case in the dataset — that is, again and again. The value of #LC is always reset to 1, so the number of times through the loop is always 5. The second time through the loop, the scratch variable #COUNT is already set to 5, so another 5 is added to it, resulting in 10 for the second line printed. The next line is 15, then 20, 25, 30, and so on, for as many cases as you have in your data.

You can write the same program with the loop counter defined separately. The built-in loop counter is named MXLOOPS (short for *maximum loops*):

```
SET MXLOOPS = 5.
LOOP.
COMPUTE #COUNT = #COUNT + 1.
END LOOP.
PRINT / #COUNT.
EXECUTE.
```

But there's a problem doing it this way: You get warning messages in your output. The purpose of MXLOOPS is as a safety measure to prevent runaway loops, so it's best to specify the count in the LOOP command. Also, either of the following methods works for defining loop termination:

```
LOOP IF (#COUNT < 5).
COMPUTE #COUNT = #COUNT + 1.
END LOOP.
```

```
LOOP.
COMPUTE #COUNT = #COUNT + 1.
END LOOP IF (#COUNT > 5).
```

BREAK

You can use the BREAK command to stop a loop. For example:

```
LOOP #LC = 1 TO 5.
   COMPUTE #COUNT = #COUNT + 1.
   DO IF (#COUNT GE 12).
      COMPUTE #COUNT = 0.
      BREAK.
   END IF.
END LOOP.
PRINT / #COUNT.
EXECUTE.
```

In this example, every time the value of #COUNT reaches 12 (or greater), the value is set back to 0 and looping stops. This program outputs 5, then 10, then 0, then 5, and so on, with one output line for every case in the dataset.

Files

You can write data to files and read data from files. Output is performed by using either SAVE or EXPORT. They keyword IMPORT can be used to read data from a file, but the simplest way to read files is to read SPSS-formatted files using GET.

GET

Whenever you choose File⇨Open⇨Data, SPSS issues a GET command to open a SPSS-formatted file and load it into SPSS. If you've loaded a file using the menu this way, you will have noticed in SPSS Viewer the GET command that loads the file. For example, the following program opens and loads the file named Cars.sav:

```
GET
  FILE='C:\Program Files\PASW\Cars.sav'.
DATASET NAME DataSet2 WINDOW=FRONT.
```

This command loads the data from the file, names it DataSet2, and opens a new SPSS main window to display the data from the file — in front of all the other windows. You never have to load a file with the menu — you can load any file from within a Syntax program by specifying its name as the first argument to a GET command.

The quotes around the filename are optional unless a blank is embedded in the name.

You don't have to load the entire contents of the file. If you want to omit certain variables, you can name them as part of the command, as in the following:

```
GET FILE="Cars.sav" /DROP=MPG DISPLACEMENT.
```

You can even change the names of some variables. For example, the following changes MPG to MILESPERGALLON:

```
GET FILE='Cars.sav' /RENAME=MPG=MILESPERGALLON.
```

IMPORT

Files saved in the SPSS portable format can be loaded into SPSS on any type of computer by using the IMPORT statement. This type of file is in a format that is portable across all computers on which SPSS runs. To read such a file into SPSS, you use the following Syntax command:

```
IMPORT FILE=DATAFILE.
```

Any files created by EXPORT (or Save As in the portable format) from SPSS on any computer can be loaded by IMPORT into SPSS on any other computer.

SAVE

The SAVE command has the same result as choosing File⇨Save As and entering a filename. It writes the data to a file in the standard SPSS format. An example of the command follows:

```
SAVE OUTFILE='C:\Program Files\PASW\Cars.sav'.
```

You have some options. You can specify DROP and RENAME the same as you can with the GET command. You can also compress the output file with the following option:

```
/COMPRESSED
```

EXPORT

The EXPORT command produces a portable data file that contains the variables and data of the current dataset. You can create such a file with a statement such as the following:

```
EXPORT OUTFILE=DATAFILE.
```

Any files created by EXPORT (or Save As in the portable format) from SPSS — on any computer — can be loaded by IMPORT into SPSS on any other computer.

Chapter 15

Command Syntax Language Examples

● ●

In This Chapter

▶ Writing a Syntax program and saving it to disk

▶ Modifying the menus to run Syntax programs

▶ Understanding some useful Syntax commands

● ●

*M*ost Syntax Command programs are short. That's because one command can do so much. This chapter is about the mechanics of writing and running Syntax programs. If you plan to do much processing with SPSS, you'll certainly be doing some things over and over. If you save those procedures in a Syntax command program, you can just run the program instead of stepping through the process again.

Writing a Syntax Command Program

To write a new Syntax program, do the following:

1. **From the main menu, choose File⇨New⇨Syntax.**

   The SPSS Syntax Editor dialog box appears, as shown in Figure 15-1. Use the text area in the upper right of the dialog box to enter the text in the Syntax language. Below the text area is a panel for displaying errors. (The error panel can be resized by dragging the bar at its top.) The panel to the left displays a synopsis of the commands in the program.

2. **Enter the text of the Syntax language in the upper-right panel.**

   The lines are numbered automatically and can be as long as you need them to be — the window will scroll as necessary.

3. **To execute the program after you write it, choose Run⇨All.**

Figure 15-1:
This is
where you
write Syntax
programs.

Syntax programs are tied tightly to the variable definitions in the currently loaded dataset. Syntax programs use the dataset's variable names, often in such a way that the type of the variable can be important. Thus the first instruction in a Syntax program is usually to load the data file.

You can load a file by choosing File⇨Open⇨Data or by writing a Syntax Command with a GET statement:

```
GET FILE='C:\Program Files\SPSSInc\
   PASWStatistics18\Samples\English\Employee data.sav'.
```

If you forget the exact form of this command, you can load a file using the menu and see the resulting Syntax command in SPSS Viewer. In fact, running *any* command by using the menu system causes its Syntax Command sequence to be written to SPSS Viewer.

The following is a program with a simple GRAPH command, using the salary and job category information of the loaded data:

```
GRAPH TITLE = "Means of Salaries"
   /SUBTITLE = "separated by job category"
   /BAR = MEAN(salary) BY jobcat.
```

The resulting display in SPSS Viewer is shown in Figure 15-2.

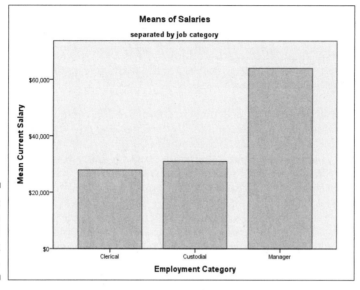

Figure 15-2:
A bar graph
produced
by a Syntax
command.

Saving and Restoring Programs

To load a Syntax program from disk, choose File⇨Load⇨Syntax from the menu of the main SPSS window. Then browse to the directory that holds the file you want to load. Select the name and click the Open button; a new Syntax Editor dialog box appears, containing the text of the program.

Having the capability to save a copy of your program is important. Whenever you write a Syntax program and think you may want to use it more than once, save it to disk so you can read it into SPSS and run it any time you want.

To save your program, first decide where you want to save it and what you want to call it. Then, in the Syntax Editor dialog box, choose File⇨Save As and choose the location and name for the new file. If you've already saved the program (or if you loaded an existing program from disk), you need only choose File⇨Save to replace the existing file.

Often you'll want to leave your original program as it is and create a new one by making changes to the original. In that case, load the original program from disk and then choose File⇨Save As to create a new file that holds your modified version. The original remains intact.

Adding a Syntax Program to the Menu

Every SPSS menu selection is nothing more than a command to execute a Syntax program. Adding a new item to the menu is a matter of adding a new menu button and assigning a task to it.

You can add new menu selections to customize SPSS and make it easier to do your common tasks. For example, if you're working on a data file and loading it regularly, you could define a new menu button to load the file for you. If you have an analysis or a report-generating program that you run regularly, you could define a menu button that runs it with your set of parameters. Or you could set up a button to export data in your preferred format.

To define terms clearly, a *menu* consists of the menu bar (the part that's always visible at the top of the window) and its row of pull-down lists. Each *list* is made up of clickable buttons. Each button can be set to execute a particular Syntax command or to display another list of buttons. You can modify a menu by adding a new pull-down list or by adding a single button to an existing list. You can delete existing menu items, but usually there's no need to do that — your modifications will almost always be for the purpose of adding buttons that perform your own tasks.

The following steps take you through the process of adding a new pull-down list with one button. The button executes the Syntax program named `loadfile.sps`, which is a program consisting of one `GET` statement to load a file:

1. **Create** `loadfile.sps`.

   Write and save a Syntax program that loads a SPSS data file. This example uses the one-line program (described earlier in this chapter) that uses the `GET` command to load `Employee data.sav`.

2. **Choose View⇨Menu Editor.**

   The dialog box shown in Figure 15-3 appears. You can make this menu choice from any system menu — the main SPSS dialog window, SPSS Viewer, or even Syntax Editor. Any menu can be modified through this dialog box.

3. **In the Apply To pull-down list, select Data Editor.**

   This is where you choose which menu you're going to modify. The other choices are View and Syntax. Each time you choose a different menu, the buttons that are already defined for that menu show up in the Menu list on the left.

4. **In the list of names in the Menu box, click the plus sign next to &Open.**

The list expands to display the items already defined: D&ata, &Syntax, &Output, and S&cript. This followed by an (End Of &Open Menu) notation. The ampersand (&) in the name specifies the following letter to be the shortcut key that activates the menu item. In this example the shortcut letters are A, S, O, and C, You can include an ampersand in the name you add, if you want.

If you use an ampersand to specify the same letter as a shortcut for more than one menu selection, SPSS will use one and ignore the other — which is probably not what you intended.

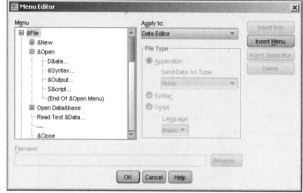

Figure 15-3:
The dialog box to add and delete menu selections.

5. **Select End Of &Open Menu.**

   The selection becomes highlighted. Whenever an item is added to the menu, it is added immediately *before* the selected item. (The End entry is included in the list only so the last position can be selected.)

6. **Click the Insert Item button.**

   A new menu button with the name New Menu Item appears.

7. **Type the name** MyFile **and press the Enter key.**

   The text you type replaces the name of the selected menu item.

8. **In the File Type area, select Syntax.**

   You can associate the new menu selection with another application or a script if you like; in this example, the new menu selection executes a Syntax program.

9. **Click the Browse button and locate the Syntax program file.**

   Clicking the button opens a browse window. Locate the directory containing the file loadfile.sps. To make the filename appear, you may need to choose Syntax Files (*.sps) in the Files Of Type pull-down list at the bottom of the dialog box.

10. **Click the OK button.**

The Open menu of the main window of SPSS has been modified by the addition of MyFile, as shown in Figure 15-4.

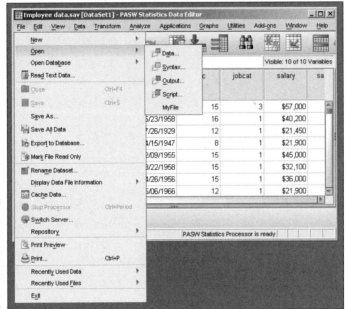

Figure 15-4:
The MyFile selection has been added to the Open menu.

If you want to add this same option to the menus of other dialog boxes in SPSS, you have to follow the same procedure for each one. Adding a menu item takes only a small amount of work and can prevent many repeated steps. For example, if you're in the process of entering and correcting data, a simple menu item to load the file would keep you from hunting for the file every time you need to load it. Also, if you have a group of analyses that you run repeatedly, you could include them all in one Syntax program and have them all run for you at the click of a button. The program could be made to load the file at its startup, so one click of a button is all you need to do all that work.

Where to Find Syntax Commands

There are lots of Syntax commands, and they all have lots of options. If you have something you want to do, and you want to find the Syntax command to do it, you have three basic approaches.

One way is to use the menu system to command SPSS to do whatever it is you would like it to do. In SPSS Viewer, you will be able to see the text of the Syntax commands that generated the output. Highlight that text, choose Edit⇨Copy, switch to the Syntax Editor dialog box, and choose Edit⇨Paste to capture the text into your own program.

Another way to find commands is to select Help⇨Topics on the SPSS menu. It may take a few tries to get the Syntax command you want, but it's listed in there somewhere. The Syntax commands are listed in all-uppercase letters, so they're easy to spot.

A third approach is to select Help⇨Command Syntax Reference on the SPSS menu. Each command is detailed, along with all its options.

Doing Several Things at Once

You can write a Syntax program to do more than one thing. All the commands in such a script are executed one after the other. And because one Syntax command can do quite a bit, you don't have to write much of a program to do lots of processing. For example, the following four-line program named `makeplot.sps` performs four separate tasks:

```
GET FILE='C:\Program Files\SPSSInc\PASWStatistics18
   \Samples\English\Cars.sav'.
DATASET NAME DataSet1 WINDOW=FRONT.
GRAPH LINE=MEAN (HORSE) BY YEAR.
GRAPH BAR=MEAN (MPG) BY ACCEL.
```

The first two lines load the SPSS data file named `Cars.sav`. The third line renames the dataset to `DataSet1` and brings the window displaying it to the front. As a result, if data has already been loaded and named `DataSet1`, this new file assumes the name (the other will be closed). The last two lines draw graphs — one line graph and one bar graph, as shown in Figures 15-5 and 15-6.

Note how variable references are made on the GRAPH commands. Referring to a variable by its name specifies that all its values be used; using the word MEAN before the variable name in parentheses specifies that the mean of the variable's values be used. These commands are simple but the actions are complex.

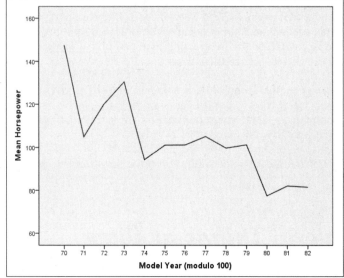

Figure 15-5:
A line graph displaying the mean horsepower for each year.

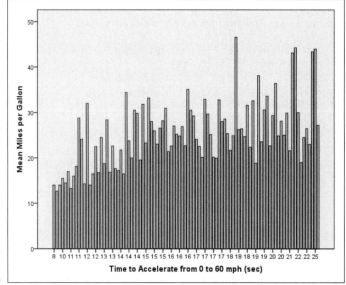

Figure 15-6:
A bar chart displaying the mean acceleration for each mile-per-gallon rating.

Graphing Q-Q and P-P Plots

The Syntax language contains the PPLOT command, which you can use to generate either of these types of plots:

✔ **q-q (quantile-quantile) plot:** The *quantiles* of the actual values are plotted against the quantiles of the expected values.

✔ **p-p (proportion-proportion) plot:** The actual proportions are plotted against the expected proportions.

The following program, named `makeplot2.sps`, contains commands to produce both.

```
GET FILE='C:\Program Files\SPSSInc\PASWStatistics18
   \Samples\English\Employee data.sav'.
DATASET NAME DataSet1 WINDOW=FRONT.
PPLOT SALARY /TYPE=Q-Q.
PPLOT SALARY /TYPE=P-P.
```

This program loads the dataset and then produces a plot of each type. The q-q plot is displayed in Figure 15-7.

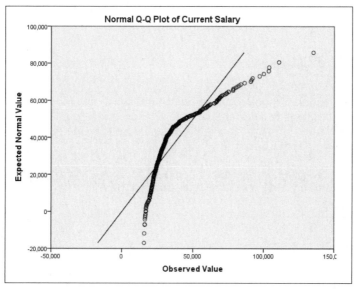

Figure 15-7: A q-q plot produced by the PPLOT command.

Figure 15-8 displays the p-p plot produced from the program.

Figures 15-7 and 15-8 don't represent all the output you get from the PPLOT command. In particular, a detrended plot (a plot in which the actual values are plotted against deviations of the expected values) is also produced.

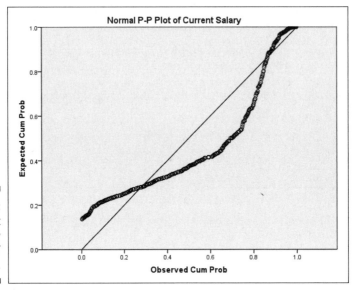

Figure 15-8:
A p-p plot
produced by
the PPLOT
command.

Splitting Cases

In this section we look at a program that loads a data file and counts the repetition of values in a certain variable. The program counts repetitions for all cases in a file, splits the file, and takes a count for each portion. The program is named splitfile.sps:

```
GET FILE='C:\Program Files\SPSSInc\PASWStatistics18
    \Samples\English\Employee data.sav'.
FREQUENCIES SALARY.
SORT CASES BY GENDER.
SPLIT FILE BY GENDER.
FREQUENCIES SALARY.
```

The first line of the program uses the GET command to load the file. The second line uses the FREQUENCIES command to generate the counts and percentages for the salary values. The top section of the table produced from this command is shown in Figure 15-9. As you can see, the table that the program generated includes five columns. A salary value is shown in the first column; a count of the total number of occurrences of the value appears in the second column. The Percent column holds the percent of the total number of cases (excluding cases with missing values in any variable) that contain this particular salary value. The Valid Percent column holds the percent of the total number of cases (including those with missing values in other variables) that contain this particular value. The Cumulative Percent is the number of cases with salaries less than or equal to the salary shown in the first column. For this example, the values displayed as Percent and Valid Percent are the same because no cases in the displayed portion contain a missing value for any variable.

Figure 15-9: A frequency table for the entire data-set.

		Frequency	Percent	Valid Percent	Cumulative Percent
Valid	$15,750	1	.2	.2	.2
	$15,900	1	.2	.2	.4
	$16,200	3	.6	.6	1.1
	$16,350	1	.2	.2	1.3
	$16,500	1	.2	.2	1.5
	$16,650	1	.2	.2	1.7
	$16,800	1	.2	.2	1.9
	$16,950	3	.6	.6	2.5
	$17,100	2	.4	.4	3.0
	$17,250	1	.2	.2	3.2
	$17,400	2	.4	.4	3.6
	$17,700	1	.2	.2	3.8

Current Salary

Then comes the split. The program that sorts cases uses the SORT command to do the split. You must sort a dataset using the same variable as the key that's about to be used to split a file; variables of like values must be all together for the split to work properly. In this example, the SORT command groups all the female cases before the male cases.

The SPLIT command logically inserts a divider at each point where the value of the named variable changes. In this example, the value f is used for female and the value m for male, so a logical divider is placed between them. The divider is logical because the split refers only to the memory-resident form of the data — the split does not survive the data being written to a file.

The last line of the program builds a new set of counts and percentages, but this time the data is divided by gender, so the table is generated in two parts. The upper part of the table is shown in Figure 15-10. The headings of the table have the same meanings they had before, but you can see that the top of the table contains the numbers for the female cases. The bottom portion contains data from males. If the SPLIT command had used a variable with more values, the cases would have been split into more parts.

Current Salary						
Gender			Frequency	Percent	Valid Percent	Cumulative Percent
Female	Valid	$15,750	1	.5	.5	.5
		$15,900	1	.5	.5	.9
		$16,200	3	1.4	1.4	2.3
		$16,350	1	.5	.5	2.8
		$16,500	1	.5	.5	3.2
		$16,650	1	.5	.5	3.7
		$16,800	1	.5	.5	4.2
		$16,950	3	1.4	1.4	5.6
		$17,100	2	.9	.9	6.5
		$17,250	1	.5	.5	6.9
		$17,400	2	.9	.9	7.9
		$17,700	1	.5	.5	8.3
		$18,150	2	.9	.9	9.3
		$18,450	1	.5	.5	9.7
		$18,750	1	.5	.5	10.2

Figure 15-10:
A portion of the separate frequency tables for females and males.

Examining Data

The EXAMINE command in the Syntax language may be the quickest way to look at data. For example, with the system data file named Cars.sav loaded into SPSS, a two-word Syntax program produces a graph of a variable. The two-word program is as follows:

```
EXAMINE MPG.
```

This command results in the box plot shown in Figure 15-11, which graphically displays the mean, the standard deviation, and the extreme values.

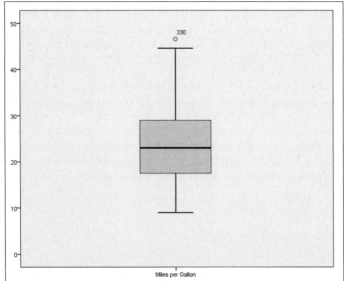

Figure 15-11:
A box plot of miles per gallon.

But that's not the only way EXAMINE can show you data. You can include more than one variable, or you can change the plot style to a histogram. The following command generates more than one histogram:

```
EXAMINE ACCEL, HORSE /PLOT=HISTOGRAM.
```

This command produces a histogram for each of the two named variables. The histogram representing the acceleration values (ACCEL) is shown in Figure 15-12.

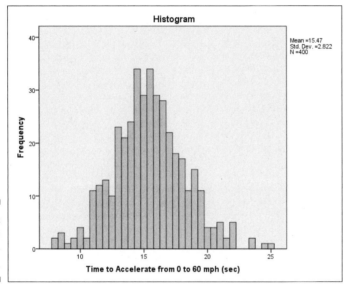

Figure 15-12:
A histogram of acceleration rates.

Part VI

Programming SPSS with Python and Scripts

The 5th Wave   By Rich Tennant

"You might want to adjust the value on your 'Nudge' function."

In this part . . .

If you've always wanted to boss a computer around in more than one language, you can write commands using the Python programming language and include them among the Command Syntax statements. The result is the same as if you had written Command Syntax, but Python is a more modern and flexible language. If you think you might want to become a computer nerd, this is the way to go.

And you can use scripting programming in either Python or Sax BASIC to create programs that execute automatically. (Careful, though this can be pretty impressive. When your friends start to bring you jelly doughnuts to crunch their numbers, be gracious.)

Chapter 16

The Python Programming Language

In This Chapter

▶ Manipulating numbers

▶ Working with strings of characters

▶ Making use of lists

▶ Uncovering the fundamental structure of a Python program

SPSS has added Python — a general-purpose programming language — that can be used as a SPSS scripting language. Chapter 17 is about using Python inside SPSS, but this chapter is about the Python programming language itself. If you're not a programmer, don't worry; Python is famous for being easy to learn. And you might think it's named after a snake, but it isn't. It's named after *Monty Python's Flying Circus*. I just thought I'd mention that in case you thought things were going to get serious. And now for something completely different.

Instructing Python: You Type It In and Python Does It

If you give Python an instruction that it understands, it will obey the instruction and do something. It's very obliging that way. But you have to be specific when you tell it what you want it to do.

If you want a Python of your own, outside of the one that comes with SPSS, you can download and install one from the Internet for free. By playing with your own Python, you can see how the examples in this chapter work. The only way to really get to know a programming language is to fiddle around with it and write some programs of your own. Sometimes you get great insight into programming from finding out what *doesn't* work.

Python is an *interpreter*. That is, instead of taking your set of program instructions and translating them to machine language, it just reads and obeys whatever you type. In effect, it reads your commands as it would read a script — so Python programs are also called *scripts*.

Python can be used to generate graphic displays, communicate over the Internet, make calls into the operating system, and some other things that we won't be messing with. This chapter shows you just enough basic Python to get you comfortable with writing scripts for SPSS.

If you want to get your own private Python, you can get your own standalone version at `www.python.org`. When you fire up the standalone version of Python, it displays >>> as a prompt for you to give it some instructions. If you type something it knows how to do, it will do it. If you type something it doesn't understand, it will complain — but it won't bite. Remember, it's not a snake.

Understanding the Way Python Does Arithmetic

Statistics is made out of arithmetic, and Python is good at arithmetic. You can enter any expression you want, and Python will do the calculations and give you the answer.

Let's start with something simple. At the prompt, type a simple addition such as the following. Python comes back with a result:

```
>>> 2 + 2
4
```

You can use multiplication, division, decimal points, parentheses, and all sorts of fancy stuff:

```
>>> (88 + 2) / 6
15
```

The symbol for multiplication is the asterisk:

```
>>> 10 * 10
100
```

If you do integers, Python does integers. If you do decimal points, Python does decimal points. Integer arithmetic just chops off, like this:

```
>>> 7/2
3
```

And arithmetic that uses decimal points (floating-point arithmetic) keeps the fractional portion in decimal form, like this:

```
>>> 7.0/2.0
3.5
```

Actually, the behavior of the decimal point in integer division varies a bit. Newer versions of Python (3.1 and later) automatically insert the decimal point for division. But it's always best to put them in just to make sure what your result will be. If you always use decimal points in your numbers, the behavior is consistent for all versions of Python.

You can mix integer and decimal numbers in the same expression, but watch what you're doing. Whenever any operation involves at least one number with a decimal point, Python treats all the numbers as if they have decimal points. For example:

```
>>> 7/2.0
3.5
```

However, you have to be really careful when you mix the number types like that. You could get something other than what you expect. The following two examples look the same, but they could be different, depending on your version of Python:

```
>>> 7.0/2.0 + 4.5
8.0
>>> 7/2 + 4.5
7.5
```

The first example performs a decimal-point division and winds up adding 3.5 to 4.5. The second example performs an integer division — which chops off the decimal part — and winds up adding 3 to 4.5. These results are different in a very practical sense: One of them is wrong for whatever you happen to be calculating.

Instead of just printing the numbers on the display, as we've done so far, you can store them in a name, called a *variable*. The three dimensions of a box could be stored in variables this way:

```
>>> height=20.0
>>> width=9.0
>>> depth=12.0
```

No number is displayed by these statements because simply storing a number somewhere doesn't display it. Python remembers those names and numbers for you. You can calculate the volume of the box and have it displayed this way:

```
>>> height * width * depth
2160.0
```

In Python, the equal sign (=) is the *assignment operator*. It takes the value from whatever you put on the right and stores it in whatever location you name on the left. It writes over whatever was there before.

If you want, you could store the volume of the box in the current example in another variable and then display it, like this:

```
>>> volume = height * width * depth
>>> volume
2160.0
```

Whatever name you enter is the one Python uses. If you spell it wrong, it's a different name, so use names that are easy to spell. And don't use things like the uppercase letter I and the lowercase letter l because they can be mistaken for each other and confused with the number 1. And watch out for the letter O and the number 0.

Python has the memory of an elephant snake. After you stick a value in a variable, Python will remember it forever. (Well, at least until you stop running the program.) If you want to really save a value, write it to a file on disk so you can read it back. (That's easy to do, and we get to it later.)

As you've seen, if you simply name a value or a variable, Python prints it for you. You can also use the print function, like this:

```
>>> print(volume)
2160.0
>>> print(height,width,depth)
20.0 9.0 12.0
```

Python remembered these variables from earlier. And you can see how the print function can handle more than one value at a time.

Understanding the Way Python Handles Words

If you want Python to notice what you're saying, you'll have to put it in quotes. You can use single or double quotes, but whichever one you use at the start is the one you must use at the finish. Like this:

```
>>> 'Single quotes'
'Single quotes'
>>> "Double quotes"
'Double quotes'
```

If you enter a quoted string by itself this way, Python echoes it back to you just as it would a number. Python usually uses single quotes when it echoes, but that's just an attitude problem and doesn't matter.

In the world of computer programming, any group of characters used to make up a name, a sentence, or anything you can read is called a *string*. Also, a blank is a character just like any other, except you can't see it if you're a mere mortal.

You can put single quotes inside double quotes and double quotes inside single quotes, like this:

```
>>> "Girl's clothes?"
"Girl's clothes?"
>>> '"Girl clothes?" he asked'
'"Girl clothes?" he asked'
```

Hmm. This time Python uses double quotes to display the string that contains a single quote. Attitude meets necessity. Python figures out which kind of quotes it needs to use to be consistent. Don't think about it too much. Let's move on to an example of storing a string in a variable:

```
>>> fred="Is this a cheese shop? "
>>> fred
'Is this a cheese shop? '
```

You can stick a string in a variable exactly the way you can a number. You can even add one string to another one, like this:

```
>>> herbie = fred + "Is this a parrot shop?"
>>> herbie
'Is this a cheese shop? Is this a parrot shop?'
```

As you can imagine, the strings can get long. You can make them show up on more than one line by inserting a \n (newline) character and using the print command, like this:

```
>>> herbie = herbie + "\nNo. This is for lumberjacks."
>>> print(herbie)
Is this a cheese shop? Is this a parrot shop?
No. This is for lumberjacks.
```

The print function translates \n as being the start of a new line. If you just echo the variable, it doesn't work — you just get a backslash and an *n* in the output.

Now for something slightly different: Using triple quotes — three sets of either single or double quotes — causes the automatic insertion of newline characters into your string whenever you start a new line. You can organize formatted text with it, like this:

```
>>> hebert="""
... Algy met a bear
... The bear was bulgy
... The bulge was Algy
... """
>>> print(hebert)
Algy met a bear
The bear was bulgy
The bulge was Algy
>>>
```

Notice that Python drops the normal >>> prompt while you're entering the triple-quoted string and uses three dots (. . .) instead. It's not important — it's just another example of Python assuming an attitude.

You can use either single quotes or double quotes to construct your triple quotes. (If that sentence makes any sense to you, you're really getting into this. Let's move on.) I showed you earlier how you can add strings; now I'll show you how they can be multiplied:

```
>>> essword="spam "
>>> print(essword * 7)
spam spam spam spam spam spam spam
```

If you want to define a long string, you can break it and enter it on more than one line, like this:

```
>>> go="Now is the time for all good men to\
... get out of town."
>>> print(go)
Now is the time for all good men to get out of town.
```

When you are entering a string of characters, you can put a backslash (\) at the end of the line and continue at the beginning of the next line just as if you had continued on the same line. As you can see by toget in the output line, I should have added a space after to and before the backward slash. You can also build long strings by adding smaller strings without putting in a plus sign.

```
>>> hank="ugly "    'dog'
>>> hank
'ugly dog'
```

You might want to do it that way and you might not. I think a plus sign between the two makes it a lot clearer, but you might want to leave it out just to show off. (That's what I was doing when I put this example in the book.)

Okay. That's enough about putting strings together. Let's take some apart. It's easy because you can refer directly to each letter by its position number. The letter at the extreme left is number 0, the next one is number 1, one after that is number 2, and so on. For example, to pull the first letter out of the string of the preceding example, you just address it by number, like this:

```
>>> hank[0]
'u'
```

If you want to extract a range of characters, just use the number of the first character you want and the number of the character *following* the last one you want, and put a colon between the two, like this:

```
>>> hank[2:6]
'ly d'
```

If you use the colon but leave out the first number, Python assumes 0. If the ending number is missing, it assumes the end of the string. Here's an example:

```
>>>> hank[:4]
'ugly'
>>> hank[5:]
'dog'
```

You can use extraction to build new strings by adding the pieces together like this:

```
>>> frank = 'very ' + hank[:4] + ' fat ' + hank[5:]
>>> frank
'very ugly fat dog'
```

One of the questions that always comes up in a program is, "How long is that string?" Here's how to find out:

```
>>> len(hank)
8
>>> len(frank)
17
```

You will find lots of functions that do things to strings that result in new strings that are different. The original string is never changed — you can't really change an existing string, no matter what you do. To make a difference in a string, you have to create a new string and then replace the original. Here are some examples of functions that create modified strings:

```
>>> hank.capitalize()
'Ugly dog'
>>> hank.find("dog")
5
>>> hank.replace('g','x')
'uxly dox'
>>> hank.title()
'Ugly Dog'
>>> hank.upper()
'UGLY DOG'
```

Remember: None of these examples changed the original. They produced new strings. But this group of functions is just the tip of the iceberg. You'll find a Python function to do just about anything you can imagine to a string

Understanding the Way Python Handles Lists

You can have a variable hold an arbitrary collection of strings and numbers. You address any specific one by its position number in the list, with the first one in the list being number 0, as in the following:

```
>>> jam=['a',100,"c",'dee']
>>> jam
['a', 100, 'c', 'dee']
>>> jam[0]
'a'
>>> jam[1]
100
>>> jam[1:3]
[100, 'c']
```

In this example, you can see where four things were stuffed into the variable named jam. When the variable was displayed, all four items it contained were displayed. You can, however, use a position value to refer to individual items in the list and address them one item at a time, or select a subset of the items in the list.

When you use a pair of position numbers, the first number is the number of the first item you want, but the second number is the number *following* the last item you want. The first item in the list is always number zero.

Position values work on lists the same way they work on strings. But strings can't be modified, only replaced. Lists *can* be modified. You can replace one member of a list by simply assigning a new thing to it, like this:

```
>>>> jam = ['a', 100, 'c', 'dee']
>>>> jam
['a', 100, 'c', 'dee']
>>>> jam[2]='hooha'
>>>> jam
['a', 100, 'hooha', 'dee']
```

You can quickly find out how many things are in a list:

```
>>> len(jam)
4
```

Lists are one of the really nice things about Python. If you want to do something to a list, try it — and it will probably work. You can even put lists inside lists, like this:

```
>>> jam[0] = ['apple', 'pear']
>>> jam
[['apple', 'pear'], 500, 'hooha', 'dee']
```

Making Functions

Python can remember a set of instructions for you, and you can later call on that set by name. Here's a simple example that divides a number in half and displays the results:

```
>>> def showhalf(x):
...     print(x/2.0)
```

The line with the def command names this as a function called showhalf. And don't forget the colon on the end of the def line. This example has one variable, named x, used in the body of the function. All statements following the definition line are included as part of the function, as long as you indent them. When you type a line that is less indented, the function ends. Python then remembers your definition of the function; you can use it as often as you like. Here's an example:

```
>>> showhalf(10)
5.0
>>> bunch=100
>>> showhalf(bunch)
50.0
```

Whatever value you include in the parentheses becomes the value of x inside the function. The following example shows that instead of just doing something inside (as in the previous example), the function can return a value to you:

```
>>> def getthird(value):
...        return(value / 3.0)
...
>>> j = 9
>>> k = getthird(j)
>>> print(k)
3.0
```

In this example, whatever value is passed to the function is divided by 3 and the result comes back because it is part of a `return` statement. You can pass anything into a function and return anything else: strings, numbers, lists, whatever you want.

It's normal to have a Python program begin with a bunch of function definitions and then have the body of the program use the functions to do its work. Functions can even call other functions, but be careful. Too much disorganization leads to *spaghetti code* — a tangle so convoluted that you can't read it.

You should know that although you can get only one value back from a function, you can pass lots of values to one. Here's an example of a function that requires more than one value for its input:

```
>>> def showsum(a,b,c):
...        print(a+b+c)
...
>>> showsum(3,5,9)
16
```

The limit of being able to return only one value from a function is never a problem. If you find that you really need to return more than one value, you can return a list, but in reality you probably need more than one function.

Here's a nifty trick: You can define your function to have some defaults for some of the values you pass to it. Then, if you leave out any of those values when you invoke the function, the defaults will be used:

```
>>> def spark(a,b="too big",c=44):
...        if (a > c):
...            print(b)
...
>>> spark(20)
>>> spark(50)
too big
>>> spark(100,"way too large")
way too large
```

In this example, the function named `spark()` has three arguments; a, b, and c. The last two have default values. The function simply tests whether the value of a is larger than the value of c; if so, it prints b. In the example, the first call to the function sets the value of a to 20, which is not larger than c, so nothing happens. In the second call, a is set to 50, and that's larger than c, so the default content of b is printed. The last call to spark() has a value for a that is larger than c, but the string printed for b is different because the value passed to the function overrides the default.

Function definitions are, in a way, the heart of the system. You normally write a program by defining your own functions and using them along with Python's plentiful built-in functions. This program structure becomes particularly convenient when you do the same sort of thing more than once, but the most important characteristic of this program structure is that you can organize your instructions in a logical way. The main problem with programs isn't writing them — it's fixing them later when they don't work the way you want. And the main problem with fixing them is finding out where to make the change. Be organized!

Making Decisions with if/else

Often you'll have a statement or two that you want to execute only under certain conditions. You can use an if statement to ask a question (essentially a true/false test); the indented statements following the if statement are executed only if the answer to your question is true. Here's an example:

```
>>> x = 3
>>> if x < 5:
...     x = 20
...     print(x)
...
20
```

You can group statements together and execute them as a single unit by putting them together as a *block* — two or more consecutive statements indented by the same amount.

Sometimes you'll want to do one thing under some circumstances and something different under other circumstances. That's where you can use else:

```
>>> x = 10
>>> if x < 8:
...     print 'x is less than 8'
... else:
...     print 'x is not less than 8'
...
x is not less than 8
```

Instead of just ending the statements in the `if` block, this code uses the `else` keyword, followed by a colon, to start a new block. Result: When the code is executed, it skips one block and runs the other. Using `if`/`else` statements this way means that one, and only one, of the two blocks of code executes.

Using `if` blocks in code is common. If you write a script of any complexity, you will nest such blocks inside one another. With a bit of practice, you will get proficient at doing such things. One odd situation comes up, how-ever — usually when you back up to change something: you find yourself having to put in some code that does nothing at all. Python is persnickety about its syntax, and there are places where you are always required to put something in, but you may find that you don't want the code to do anything at that point. To the rescue comes the keyword `pass`, which you can use like this:

```
>>> x = 3
>>> if x < 8:
...      pass
... else:
...      print('x is not less than 8')
...
```

This example has no output; all it does is execute the `pass` command, which does absolutely nothing. But perhaps you want to use the `if` statement to select a single action among several possible choices. You can do that as fol-lows:

```
>>> x = 8
>>> if x < 8:
...      print 'x is less than eight'
... elif x == 8:
...      print 'x is equal to eight'
... else:
...      print 'x is greater than eight'
...
x is equal to eight
```

The `elif` keyword is short for *else if* — you can use it to add another condition, followed by another block of statements that will be executed only if that second expression is true. You can daisy-chain as many of these `elif` statements as you want, and only the first one found to be true is executed — the rest are skipped. You can have only one `else` statement, and it must come last.

When you're telling the computer what to do, be sure to say what you mean. Although the single equal sign (=) is the assignment operator and is used to copy data, the test for a couple of values being equal is the double equal sign (==). You can include the *greater than or equal to* test with >=, the *less than or equal to* test with <=, and the *not equal to* test with !=.

You can also use and, or, and parentheses in expressions. Things can get complicated if you need to ask a hard question. For example, the following is true only if aa is greater than or equal to bb *and* x is not equal to y:

```
>>> if (aa >= bb) and (x != y):
```

Don't think too much about what that statement means. (That just leads to headaches.) I wanted to show it to you so you'd know that that sort of thing is possible if you really need it, or if you find yourself with a sudden urge to do something baroque.

Doing It Over Again with for and while

Repetition occurs often in programming because it's often necessary. Having your program go back through the same code again is called *looping,* or *iteration.* (You're probably familiar with the word *reiterate,* which means to repeat something.)

You can iterate in Python by using the for keyword, like this:

```
>>> bog = ['first',50,'third',800,3.14159]
>>> for x in bog:
...     print(x)
...
first
50
third
800
3.14159
```

First you create a list and then set up a variable in the for loop to iterate through the list. The loop executes once for each member of the list; for each iteration, the variable assumes the value of a member of the list. It couldn't be easier. (Well, if you think of an easier way, tell the folks at Python, and I'm sure they'll put it in the language.)

Don't change any of the values in the list while you're inside the loop. The results of doing that are unpredictable, and the last thing you want in your computer is a confused Python. If you absolutely, positively *have* to change the list inside the loop, use a copy of the list to iterate.

It's common in other programming languages to iterate a specific number of times. You can do that in Python if you feel you must. A special, built-in function called range() returns a list and lets you iterate a set number of times. You can do it this way if you feel an irresistible urge to count:

```
>>> for z in range(5):
...      print(z)
...
0
1
2
3
4
```

Or you can use the range() function for starting at some value other than 0, like this:

```
>>> for y in range(5,10):
...      print(y)
...
5
6
7
8
9
```

Iterating by a count is actually not a different capability of the language — the range() function simply returns a list containing the numbers needed for the count. But a different capability of the language *is* found in the other iterater, named while. It works a lot like if, except it repeats continuously, testing a conditional expression to determine when to stop. while continues to execute its block of statements as long as the condition it tests comes up true. The following is a simple example:

```
>>> x = 2
>>> while x < 8:
...      print(x)
...      x = x + 1
...
2
3
4
5
6
7
```

When inside a while loop, make sure you do something that affects the value of the expression tested by the while command. Otherwise, you could be caught in the loop forever. And that's an embarrassingly long time.

I said earlier that a while statement is sort of like an if statement. In fact, it is so much like an if statement that you can put an else at the end of the block of a while statement, like this:

```
>>> x = 7
>>> while x < 9:
...     print(x)
...     x = x + 1
... else:
...     print("The loop is done")
...
7
8
The loop is done
```

The first part of the loop works just like an if statement, except it executes over and over as long as the conditional expression is true. When the expression becomes false, the else part of the statement executes once — and then the while statement is finished.

"But, hold varlet," you shout, drawing your sword. "A statement following the loop would execute once without regard to the presence of else." Whereupon I wisely retort, "Stay your hand. Bear with my discourse but a bit longer and I will show you purpose." Then I cleverly explain the operations of continue and break.

A continue statement anywhere inside a for loop or a while loop will cause the rest of the statements inside the loop to be skipped. That is, the continue keyword jumps immediately to the bottom of the loop, allowing things to come back around again normally.

A break statement inside a loop will cause the while or for loop to be abandoned as if all iterations had completed, regardless of whether that is the case. In fact, when a break statement abandons the execution of a loop, it will also cause any terminating else code *to be skipped.* This is where you slip your sword back into its scabbard, muttering, "I'll get you next time."

"One more thing!" I shout. "It is common to nest for and while loops inside one another. When that happens, the continue and break statements only continue or break the innermost loop." I mention this only because it's the kind of thing that can send you on a long and fruitless bug hunt.

Chapter 17

Python inside SPSS

· ·

· ·

*T*his chapter is the gateway to becoming a SPSS power user. It includes the mechanics you need to know to be able to write Python programs that run inside SPSS. Integrating Python with SPSS makes possible some things that would otherwise be difficult to do in the Syntax language. To use Python, you need to know the basics of the SPSS Syntax command language because you actually reach out of Python and into the Syntax language to issue commands to SPSS. You can think of the Python plug-in as an extension of the built-in Syntax language.

Python was designed to be a general-purpose language, so it has a much larger scope than you'll ever need for SPSS programming. This large scope means that it contains features and capabilities you'll never use within SPSS. On the other hand, it also means that you can solve special problems unique to your situation.

Don't let Python's size intimidate you. It's sort of like having a pocket calculator with lots of extras — if you see a button that doesn't make sense to you, ignore it.

Installing Python for SPSS

Python is not installed as part of the SPSS base system; you have to install it separately. Whether you got your SPSS from a CD or downloaded it from the SPSS Web site, the process of installing Python is fundamentally the same:

1. **Install the IBM SPSS Statistics Python Essentials Package.**

   The package is provided with your IBM SPSS Statistics product. It includes the Python Integration plug-in. When you first execute the installation process, the window in Figure 17-1 appears. For SPSS version 18, the Python version is 2.6. (SPSS versions 16 and 17 use Python version 2.5; SPSS Versions 14 and 15 use Python version 2.4.) Version is more important with Python than with other languages because they are not compatible with each other.

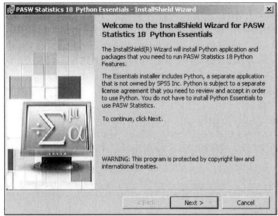

Figure 17-1:
The Python Essentials package is not essential to the operation of SPSS.

2. **Click the Next button.**

   The license agreement appears, as shown in Figure 17-2.

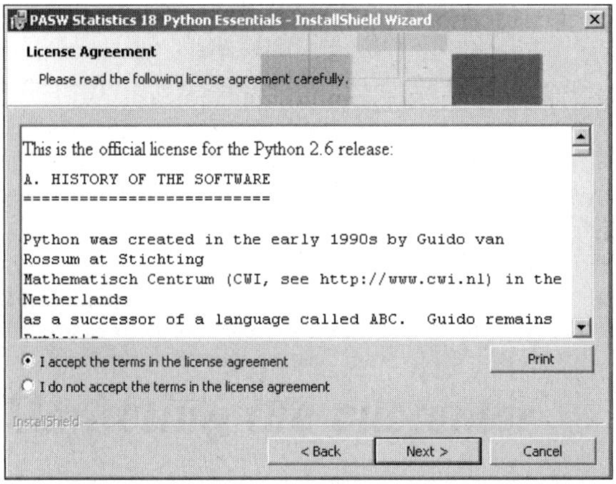

Figure 17-2:
The Python 2.6 license agreement.

3. **If you decide to continue after reading the license, click the I Accept the Terms option and then click the Next button.**

You should read the license agreement before you accept it because you'll be bound by it.

4. **A second license window will appear — the freeware license. Again, to continue, accept the terms and click the Next button.**

These terms are very easy to accept. At any rate, you can cancel them at any time by deleting the software.

5. **Yet another window appears giving you the opportunity of canceling Python installation. Click the Install button.**

This window names the exact version of Python being installed, and the two modules: Python Application and IBM SPSS Statistics Python Plug-in.

6. **Choose whether you want to allow Python to be accessible to other logins on the computer (see Figure 17-3) and then click Next.**

Install Python for all users unless you have a specific reason to exclude someone.

Figure 17-3: Python can be limited to a single user or it can be open to all users.

7. **Select the name of the directory to contain the Python files, as shown in Figure 17-4, and then click Next.**

Although you can choose any directory name and disk on your system, I suggest you use the default directory, `Python26`. If you do use a different directory, don't use an existing one with other files already in it.

8. **Decide whether to leave out some of the Python files.**

Unless you have a specific disk-space problem, accept the default, as shown in Figure 17-5, and install everything. If you have limited disk space, click Disk Usage to see how much space you have (and which disks have enough space). You can click the Back button to change the original location of your installation.

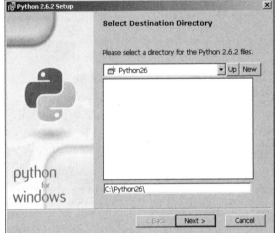

Figure 17-4:
You can install Python in any directory on your system.

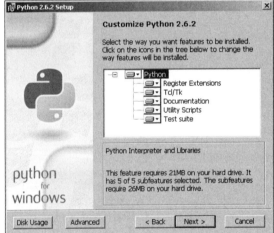

Figure 17-5:
You can exclude parts of Python from the initial installation.

9. Click Next.

A dialog box with a progress bar appears, and the bar moves all the way across a few times. The progress bar vanishes and a Finish window appears.

Depending on what has been previously installed, you may be presented with the Install button again. If so, you will need to click the Install button and repeat some of the above steps for the plug-in portion of the software.

10. Click Finish

You will actually need to click the Finish button more than once. The Finish window vanishes, leaving the original Install window in place. The Python language is now installed. You can now use that software to execute Python programs.

Python and the SPSS Python plug-in are now installed and ready to go to work for you.

Using a Language inside a Language

Python runs from inside a Syntax program. After you have the Python plug-in installed, you can write Syntax programs and include Python programs inside them by surrounding the Python code with the correct Syntax commands.

To run a simple Python command, and to check whether your plug-in is installed and working, do the following. In the main SPSS window, choose File⇨New⇨Syntax. Then, in the panel on the right in the Syntax Editor dialog box, enter the following three lines:

```
BEGIN PROGRAM.
print "Python speaks!"
END PROGRAM.
```

When you enter the program, it looks like the one in Figure 17-6.

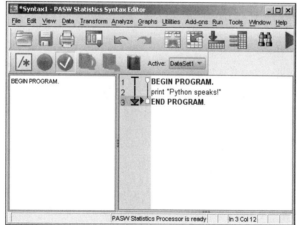

Figure 17-6: A simple one-line Python program.

Don't forget the periods following the two Syntax language directives. They are tiny, hard to see, and easy to forget. But the thing won't work without them.

This is a Syntax program with a one-line Python program embedded inside it. Everything between the Syntax commands BEGIN PROGRAM and END PROGRAM is Python. In this example, the included program is one line of Python consisting of a print statement. Choosing Run⇨All on the Syntax Editor menu produces the following text as output:

```
BEGIN PROGRAM.
print "Python speaks!"
END PROGRAM.
Python speaks!
```

The display consists of a complete listing of the program, followed by the output from the program. I'm sure you've notice from your earlier activities that SPSS always lists the Syntax source code before running the program.

Python runs inside SPSS, but you have to tell Python where it can find things. To access SPSS data and commands from inside a Python program, you must use an import statement to make data from SPSS available. You import a package such as spss only once in a Python program, but after you do, you have access to all SPSS data and even to Syntax language commands. (The spss package contains links to the Syntax commands and paths to the data-set.) For example, from inside a Python program, you can use the Syntax command LIST to output the values of all the variables of all the cases:

```
BEGIN PROGRAM.
import spss
spss.Submit("LIST.")
END PROGRAM.
```

To be able to run this and other Python programs, it is necessary to have a dataset loaded into SPSS. Without data, the program will generate error number 105 and no useful output.

This is a Syntax command issued from inside a Python program which is, in turn, being run from inside a Syntax program. I need to mention something you should be careful about here: Notice that all the Python code is in low-ercase, except for the capital *S* in Submit. That's because Python is case-sensitive; getting the case wrong is the same as misspelling a word. Syntax language, on the other hand, doesn't care about case.

Finding out about modules

You can import Python modules that have classes and functions in them to help you with processing. In the examples in the preceding section, the spss module contains the Submit function, which is used to execute Syntax com-mands. The spss module also contains other useful functions. You can look at the name of all its functions with the following program:

```
BEGIN PROGRAM.
import spss
help(spss)
END PROGRAM.
```

This program uses the Python help function to output information about the module. But the information is larger than can be displayed in SPSS Viewer in a single chunk, so it's necessary to expand things to see it all. Double-click the text shown in SPSS Viewer, and the complete text (all 1769 lines of it) appears in an SPSS Text Output dialog box. You can scroll through the text to find out what's there.

You can use help to find out about almost anything having to do with Python. For example, if you want to be more specific in your search for help, you can get help on the Submit function by executing the command help(spss.Submit). You can also be more general in your search. For example, the following program gives you a complete list of available modules:

```
BEGIN PROGRAM.
help("modules")
END PROGRAM.
```

You'll find that there are lots of modules — more than you'll ever use. And you can import as many as you need simultaneously.

To find out about the contents of a specific module, import it and then list documentation on its contents. For example, the following program produces a list of functions available in the time module:

```
BEGIN PROGRAM.
import time
help(time)
END PROGRAM.
```

You can even get help on help with the following:

```
BEGIN PROGRAM.
help()
END PROGRAM.
```

Installing more modules

A large number of modules are installed with your SPSS Python installation, but even more are available. And you can find information about how to install and use them on the SPSS Web site at the following address:

```
www.spss.com/devcentral/
```

You'll find several Python modules at this site, but you probably already have everything you need. If you get an urge to do lots more things with Python, I suggest taking a look at the documentation for the following modules:

- ✔ `spss`: This is the fundamental SPSS access module discussed in the previous section.

- ✔ `spssaux`: This module contains utilities, many of which are used by other modules. Among other things, you can use these utilities to work with SPSS definitions and produce output. `spssaux` provides pathways for data coming out of SPSS to be input into Python.

- ✔ `spssdata`: This module provides access to the data of the current SPSS dataset. It can be instructed to fetch the data one case at a time or loop through all the cases that return data to your program.

Executing Multiple Commands with One Submit

You can use the `Submit` function to execute more than one Syntax statement. You can either use a series of `Submit` statements, or you can issue a series of statements in a single `Submit` function call. The following example shows you how you can use the `Submit` function call with an array of command strings:

```
BEGIN PROGRAM.
import spss
spss.Submit(
   ["GET FILE='c:/Program
            Files/SPSSInc/PASWStatistics18/
            Samples/English/Cars.sav'.",
    "PRINT / ALL.",
    "EXECUTE."]
)
END PROGRAM.
```

To print this example, I had to split the quoted string, but in the SPSS Syntax editor window, it can be entered as a single line. With this form of the call to `Submit`, all the punctuation must be correct so Python can figure out what you mean. Pay special attention to these details:

- ✔ The square brackets ([and]) indicate an array of quoted strings instead of a single quoted string.

- ✔ Each string inside the array has its own beginning and ending double quotes (") to delimit the beginning and ending of the string.

- ✔ Use forward slashes (/) to construct the path name of a file. Backward slashes won't work because they have a special meaning to Python.

✔ A string within a string is delimited by using a different kind of quote. In this example, single quotes (') define an inside string and double quotes (") define the containing string.

✔ Be sure to place a comma between the strings of the array. Without that comma, Python combines the strings into one.

✔ Terminate every Syntax command with a period.

Working with SPSS Variables

You can read the values of SPSS variables and do an analysis on them inside a Python program. The PASW module gives you access to them. The following example does a simple analysis by using only the scale variables:

```
BEGIN PROGRAM.
import spss
spss.Submit("GET FILE='c:/Program Files/SPSSInc/
        PASWStatistics18/Samples/English/Cars.sav'.")
varList=[]
for i in range(spss.GetVariableCount()):
    if(spss.GetVariableMeasurementLevel(i)=='scale'):
        varList.append(spss.GetVariableName(i))
if(len(varList)):
    spss.Submit("DESCRIPTIVES " + " ".join(varList) + ".")
END PROGRAM.
```

This example program performs the following actions:

✔ The spss module is imported.

✔ The spss.Submit function is called to load the information from disk.

✔ An array named varList is declared. The array is initially empty.

✔ A loop executes, in which the variable i ranges from 0 to the total number of variables in the loaded dataset. The total number of variables is determined at the top of the loop by a call to GetVariableCount().

✔ Inside the loop, the call to GetVariableMeasurementLevel() returns a descriptor of the variable's type. If it's a scale type, the variable name is retrieved with a call to GetVariableName(), and then the name is appended to the array varList[].

✔ Inside a second if statement, a call to len() determines whether anything has been added to the array. If nothing has been added, then there are no scale variables, and no output is produced.

✔ If at least one variable is in the array, a call to Submit() executes a Syntax language DESCRIPTIVES command. The result is the output shown in Figure 17-7.

Figure 17-7:
A table
produced
by Python's
execution
of a Syntax
DESCRIP-
TIVES
command.

Descriptive Statistics					
	N	Minimum	Maximum	Mean	Std. Deviation
Miles per Gallon	398	9	47	23.51	7.816
Engine Displacement (cu. inches)	406	4	455	194.04	105.207
Horsepower	400	46	230	104.83	38.522
Vehicle Weight (lbs.)	406	732	5140	2969.56	849.827
Time to Accelerate from 0 to 60 mph (sec)	406	8	25	15.50	2.821
Valid N (listwise)	392				

In this example, a group of variable names were entered as part of a single string. The command string looked like this:

```
DESCRIPTIVES mpg engine horse weight accel.
```

Here the join() method is a Python method that accepts an array of strings and joins them as one long string, with spaces inserted as separators.

Accessing SPSS from Outside

You don't have to load SPSS and use the Syntax window to run Python. Python runs on its own — and you can use SPSS commands within the stand-alone Python program. The two magic words are

```
import spss
```

From the import statement in your Python programs, you can call Submit() or any other function defined in the spss package. You can also load other packages as you need them. You don't have to use BEGIN PROGRAM and END PROGRAM because you don't have to issue a notification of your intent to use Python.

An IDE (Integrated Development Environment) for Python provides you with everything you need to get Python to work with Syntax. A Python IDE has a built-in text editor designed for the Python language, a Python runtime system, a debugger, and the capability to load modules. Several IDEs exist — use Google to search for *Python IDE* and you'll find several. You'll like some better than others, so don't just hang out with the first one you come across.

Chapter 18

Scripts

● ●

In This Chapter

▶ Scripting with Sax BASIC for SPSS

▶ Examining BASIC classes and objects for SPSS

▶ Creating global and automatic scripts

● ●

*Y*ou can write BASIC language programs that run inside SPSS. Such programs are known to SPSS as *scripts*. SPSS has a dialog box specially designed for editing these scripts, running them, and saving them to disk. When you write scripts, you have the advantage that the Sax BASIC language is common and widespread — making it easy to find documentation, both in print form and on the Internet. A good deal of documentation is also available inside the SPSS help system.

Although scripts can be made to work with input data, they primarily work with output data — the data displayed in SPSS Viewer. For example, you can use a script to add items to or delete items from a pivot table. Also, you can write a script to modify a graph after it has been displayed.

Picking Up BASIC

This chapter is not a tutorial on programming using the BASIC language. (You can get that information from Internet tutorials and from books on Sax BASIC and Visual BASIC.) This chapter is about the particulars of using BASIC as a scripting language inside SPSS.

You should always start writing a new script by copying an old script that works. SPSS provides a number of starter scripts for you to use for this very purpose. Before you write a script of your own, look through this collection and get familiar with the available scripts. They're complete, working scripts; one of them may already perform the task you're trying to do.

The following Web site is an excellent source for examples. They are organized into categories, and you may find the exact script you're looking for. If not, you can get one that is similar and modify it to do what you want:

```
http://pages.infinit.net/rlevesqu/SampleScripts.htm
```

Scripting can be used to automate lots of things, but it does not provide magic powers for you to do things you cannot do manually. All the things you can do with a script, you can also do with mouse controls. Before writing a script, you should step through the procedure with the mouse so you know exactly what you want the script to do.

In the SPSS system, only both BASIC and Python programs are referred to as scripts. Although the Syntax language fits the technical definition of a scripting language, SPSS considers BASIC as its primary scripting language and Python as an optional add-on scripting language.

Scripting Fundamentals

The dialect of the BASIC language used in SPSS is common and known generally as Sax BASIC. Sax BASIC uses a few of the fundamental concepts of object-oriented programming. It doesn't use many, but you need to have an understanding of the little bits it does use.

Through some process that I don't quite get, object-oriented programming has gained the reputation of being difficult to understand. It isn't. It's easy to understand, but it is clumsy to explain — it's sort of like describing an accordion without using your hands. But let me try.

The roads and streets are full of cars. There are many different kinds and shapes of cars, but they are all cars. That means the word *car* is a specific classification of vehicle. A car, then, is a *class*. Fred's old, beat-up, blue 1968 Chevy is a specific car. It is an *object* of the class known as car. Every actual car is an object.

In the preceding paragraph I made reference to Fred's car. It was only a reference; not the actual object.

If you have all that — class, object, and reference — you now understand every fundamental that you need to be able to understand object-oriented programming. If you find yourself getting confused about which is what, just remember Fred's old, beat-up, blue Chevy. That's what I do, and it works for me.

Software classes, objects, and references

You already know what a pivot table is (if not, it's in the glossary). And you know that although lots of pivot tables of different sizes and types exist, they're all pivot tables. That makes a pivot table a classification — or, in programming terms, a *class*. A specific pivot table is an *object*.

In SPSS scripts, a pivot table is an object of the class named `PivotTable`. You can't copy an entire pivot table into your program, but you can get a *reference* to it. Think of the reference as a kind of address that provides access to the pivot table when you want to refer to it. In your script, you can create a reference to a pivot table with a statement like the following:

```
Dim pt as PivotTable
```

In this statement, the `pt` variable is created as a reference to an object of the `PivotTable` class. The class name, `PivotTable`, is already defined for you by SPSS. Class names are already defined for charts, documents, data cells, and several other things. (You can find a complete list in the next section as Table 18-1.) The reason the silly word `Dim` is used to declare a variable has to do with the boring history of the BASIC language. I chose `pt` to be the name of the reference for no particular reason; you can choose any name you like. The names used for references in the example programs supplied by SPSS are made by sticking `obj` in front of the class name, as in the following:

```
Dim objPivotTable as PivotTable
```

A reference is not an object, but the only thing it can ever do is *refer* to an object, so the name of a reference beginning with "obj" should not be completely misleading.

A new reference declared this way does *not* refer to an actual pivot table object. Yet. For it to do so, you have to select a `PivotTable` object and initialize your new reference variable with its address.

A few classes are built into Sax BASIC; you encounter them in the sample scripts. For example, a class named `String` is used to declare string variable references such as the following:

```
Dim mystring as String
```

Or you can define the reference to an `Integer` like the following:

```
Dim myinteger as Integer
```

The classes of SPSS

A number of classes are defined and ready for you to use in your program, as listed in Table 18-1. All names of all classes (with the exception of `PivotTable`) begin with an uppercase *I.* All the references in the example programs begin with lowercase letters.

One member of the list is special: The reference name `objSpssApp`, which is of the class `ISpssApp`, has already been declared and initialized. It's ready to go in every program; it acts as your access point to objects in all the other classes. By using the properties and methods of `objSpssApp`, you can acquire objects in all the other classes.

Table 18-1	Predefined Classes You Can Use in Your Program	
Class Name	*What an Object of This Class Refers To*	*Name Used in the Example Programs for References*
PivotTable	Pivot table	objPivotTable
ISpssApp	Entire SPSS application	objSpssApp
ISpssChart	Chart or graph	objSPSSChart
ISpssDataCells	Data cells	objDataCells
ISpssDataDoc	Data document	objDataDoc
ISpssDimension	Dimension	objDimension
ISpssDocuments	Documents	objDocuments
ISpssFootnotes	Footnotes	objFootnotes
ISpssInfo	SPSS file information	objSpssInfo
ISpssItem	Output item	objOutputItem
ISpssItems	Collection of output items	objOutputItems
ISpssLabels	Row or column labels	objColumnLabels and objRowLabels
ISpssLayerLabels	Layer labels	objLayerLabels
ISpssOptions	SPSS options	objSpssOptions
ISpssOutputDoc	Viewer document	objOutputDoc
ISpssPrintOptions	Printer options	objPrintOptions
ISpssPivotMgr	Pivot manager	objPivotMgr
ISpssRtf	Text	objSPSSText
ISpssSyntaxDoc	Syntax document	objSyntaxDoc

Properties and methods

Each class has a unique set of properties and methods by which you can access its internal information. A *property* is a variable that is part of the class definition. Each object of a class has its own set of values for its properties. Each property has read and write permission settings. Your program can use some properties only to read values from the object, other properties to write values into the object, and still other properties for both. *Methods* are procedures associated with the object, making it possible for you to execute a set of instructions associated with the object.

To be able to do anything with an object, you need to know which properties and methods are available. You can find out about any particular class definition by following these steps:

1. **Choose File⇨New⇨Script**

    This opens the dialog box used to edit scripts.

2. **Choose Help⇨IBM SPSS Statistics Object Help.**

    The dialog box shown in Figure 18-1 appears, showing you the relationship among classes as clickable buttons.

    Some versions of the Vista operating system will get an error with this operation. A patch should be available from Microsoft.

3. **Click the button representing the class you want to know about.**

    A window appears with a brief description of the class. This same window may contain other information, such as example code that shows how to declare a reference and how to initialize the reference with a specific object.

4. **Click Properties or Methods button to get more information.**

    You are presented with a list of either property or method names.

5. **Select a name from the list and then click Display.**

    A full description of the property or method appears, along with the syntax of the code you can use to access it.

In the class-name layout shown in Figure 18-1, you can see the relationships among the classes. Every class (except the Application class at the very top) is derived from another class. It's more of that object-oriented programming stuff. These relationships are important. For example, the OutputItem class is a special version of the OutputDocument class. That is, an OutputItem *is* an OutputDocument, but with some special features added. They've devised some special object-oriented words for all this ("polymorphism" and "inheritance"), but that's all just nerd mouthwash.

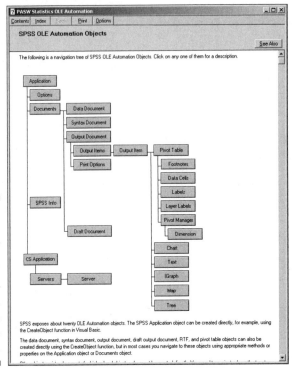

Figure 18-1:
A graphic
display
of SPSS
classes and
their
relation-
ships
to one
another.

Creating a New Script

The first step in creating a script is to choose File⇨New⇨Script. A dialog box appears with the seed of a script, as shown in Figure 18-2. The top line is `'#Language "WBB-COM"`, which is required for the program to be Visual Basic compatible. The `Option Explicit` line is needed to require that all variable names be explicitly declared using `Dim` statements — this is something you want, otherwise a simple misspelling could create an unreported error that would prevent your program from running. The seed of the executable portion of the program consists only of the opening line `Sub Main` and the closing line `End Sub`.

But this skeletal start is not all the help that's available, nor is it all the help you *should* want. The BASIC programming language is a bit strange; my advice to you is to start with a working example. You can find examples in several places on the Web, such as the location mentioned earlier in this chapter. You will also find some examples that were installed with your SPSS software. If you used the default installation directory, you can find them in the following location:

```
c:\Program Files\SPSSInc\PASWStatistics18\Samples
```

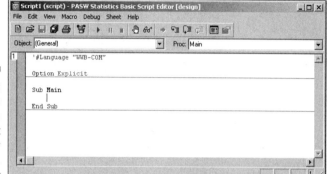

Figure 18-2:
The
skeleton
start script
provided by
SPSS.

You may have to browse around a bit to find the script that's most like the one you want to produce. These are not tiny scripts; each one has several lines of code and is filled with comments explaining how it works and how you might want to change it to make it do what you'd like.

After you have found a script you want to use for your starter, save it immediately under a new name. You don't want to save it under the same name because your changed script will overwrite the starter script — and then you won't be able to get the starter back if you need it.

While you are editing a script, you need to save it to disk from time to time for safety, and then save it again when you're finished. Your script file can be stored anywhere, but it should have the suffix .wwd (or .WWD, case doesn't matter) so you can load it into SPSS and use it again.

Automatic Scripts

You can configure scripts to execute automatically whenever output is created. There are two ways to do it. You can schedule a script to run triggered by all output, and you can schedule scripts to run triggered by specific types of output. Both ways are scheduled by choosing Edit⇨Options which brings up the dialog box shown in Figure 18-3.

First, you can set a single script to execute automatically on all data output. To do this, enter the name of the script file as the Base Autoscript in the dialog. This script will execute before any other automatic script scheduled for the data.

Second, you can set scripts to execute automatically on specific types of output from specific commands. IBM SPSS Statistics has hundreds of discrete triggers, and each one can be assigned its own script. You do it by using the scrollable panels at the bottom of the dialog box in Figure 18-3.

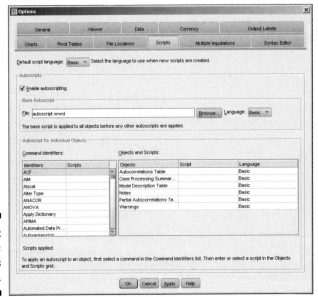

Figure 18-3:
Automatic
scripting is
enabled.

Syntax Language statements control every action that SPSS takes. You can use the panel on the left to select a Syntax Language command; you can (optionally) assign a script to that command and have the script run whenever the command is executed. Every time you make a selection in the panel on the left, the panel on the right becomes a list of specific actions that can result from running that command; you can choose the specific action and attach a script to it.

Part VII
The Part of Tens

The 5th Wave   By Rich Tennant

"Our automated response policy to a large company-wide data crash is to notify management, back up existing data and sell 90% of my shares in the company."

In this part . . .

This part describes directions you can go to explore beyond the contents of this book. You will find ten modules that can be added on to IBM SPSS Statistics, and ten places you can go on the Internet to find useful information. Go forth now and crunch.

Chapter 19

Ten (or So) Modules You Can Add to SPSS

*I*BM SPSS Statistics comes in the form of a base system, but you can acquire additional modules to add on to that system. If you have installed a full system, you may already have some of these add-ons. Most are integrated and look like integral parts of the base system. Some may be of no interest to you; others could become indispensable. Dozens of add-on modules are available. This chapter introduces you to some of the modules that can be added to SPSS and what they do; refer to the documentation that comes with each one for a full tutorial.

All the add-ons listed here except one come directly from SPSS, and you can find out more about them at the IBM SPSS Statistics Web site (www.spss.com). All add-ons are available in English. Some are also available in Japanese, French, German, Italian, Spanish, Chinese, Polish, Korean, or Russian.

Amos

Amos is an interactive interface you can use to build structural equation models. Using the diagrams you create with Amos, you can uncover otherwise-hidden relationships and observe graphically how changes in certain values affect other values. You can create a model on non-numeric data without having to assign numerical scores to the data. You can analyze censored data without having to make assumptions beyond normality.

Amos provides a more intuitive interface than plain SPSS for a certain family of problems. Amos contains structural modeling software that you control with a drag-and-drop interface. Because the interface is intuitive, you can create models that come closer to the real world than the multivariate statistical methods of SPSS. You set up your variables and then you can perform analyses using hypothetical relationships.

Amos enables you to build models that more realistically reflect complex relationships with the ability to use observed variables, such as survey data or latent variables like "satisfaction" to predict any other numeric variable. *Structural equation modeling*, sometimes called *path analysis*, helps you gain additional insight into causal models and the strength of variable relationships.

Direct Marketing

The Direct Marketing module offers a set of operations that are designed to make your direct marketing campaign more effective. You can use them to identify demographics, determine purchasing patterns, and find other useful characteristics that may be hidden inside your data. You can define and target specific groups of potential customers and maximize positive response rates.

It consists of six specific operations:

- **RFM Analysis** considers the recency of a purchase, the frequency with which that customer purchases, and the monetary amounts involved.
- **Cluster Analysis** explores your data and uncovers natural groupings.
- **Prospect Profiles** looks at your previous campaigns to generate profiles of responders.
- **Postal Code Response Rates** compares the results from various zip codes.

✓ **Propensity to Purchase** scores your database with likely-to-respond scores.

✓ **Control Package Test** compares the results of marketing campaigns.

SPSS Missing Values

The missing values in your data (whether or not they are excluded from your calculations) have an effect on the outcome. The SPSS Missing Value Analysis add-on can let you know in what way the pattern of missing values is affecting your results.

With this add-on, you can detect patterns of missing data. Armed with this information, you can determine the cause of the missing information or you can use regression or expectation algorithms to generate values. By properly managing missing data, you can use *all* your data instead of limiting analysis to complete cases. Handling missing data wisely can remove hidden bias.

SPSS Missing Value Analysis can tell you whether you have a serious missing data problem. You can find this out through the *data-patterns report*, which is a case-by-case overview displaying the extent and overview of the missing data.

The missing data information can be used to improve survey questions that you identify as possibly troublesome or confusing. You can detect whether there is a relationship between missing values — values missing for one variable could be related to missing variables of another.

SPSS Data Collection Data Entry

This is a combination of SPSS Data Collection Author and SPSS Data Collection Interviewer. This combination makes it possible for you to create professional surveys quickly and then capture data efficiently.

This module supports the construction of all types of questions — single response, multiple response, text, date, numeric, and complex questions about matrices and grids. You can use your existing SPSS data files to construct your questions.

During data collection, sophisticated validation techniques, including go-to, list filtering, and skip-and-fill, ensure as much accuracy as possible. Accuracy is increased through the use of a configurable keyboard interface along with visual and audible cues. There are also facilities for multiple data-entry operators.

SPSS Regression

The SPSS Regression add-on plugs directly into the SPSS base software and provides a larger selection of statistical analysis methods. It includes some additional diagnostic capabilities. With it, you can predict group membership in key groups. You could, for example, build a model that predicts which product a customer is most likely to order.

With this add-on, you can regress a categorical variable with multiple categories based on a set of independent variables. You can analyze data by using statistical techniques such as forward entry and backward elimination, and move in steps forward or backward (a process that exposes the most accurate predictors). If you wind up with a large number of predictor variables, you can use them together to come up with a more accurate result. Using this add-on, you can group people according to their predicted actions.

SPSS Advanced Statistics

The SPSS Advanced Statistics add-on specializes in complex relationships among multiple variables. The procedures are more sophisticated than the multivariate algorithms found in the base SPSS system; you can use them to produce more dependable conclusions. This is a set of univariate and multivariate analysis techniques that you can apply to real-world problems.

In addition to the general linear models and mixed models, SPSS Advanced Models now includes procedures for Generalized Linear Models (GLMs) and Generalized Estimating Equations (GEEs).

The GLMs include linear regression for normally distributed responses, logistic models for binary data, and log-linear models for count data. The GEEs extend linear models to accommodate correlated longitudinal data and clustered data. The SPSS Advanced Model also includes GLM and Hierarchical Linear Models (HLM).

Statistical terms describe types of calculations. You will find descriptions of them in the glossary.

SPSS Exact Tests

The SPSS Exact Tests add-on makes it possible to be more accurate in your analysis of small datasets and datasets that contain rare occurrences. It gives you the tools you need for analyzing such data conditions with more accuracy than would otherwise be possible.

When only a small sample size is available, you can use this add-on to analyze that smaller sample and have more confidence in the results. Here the idea is to perform more analyses in a shorter period of time. This add-on allows you to conduct different surveys rather than spend time gathering samples to enlarge the base of the surveys you have.

The processes you use, and the forms of the results, are the same as those in the base SPSS system, but the internal algorithms are tuned to work with smaller datasets. The Exact Tests add-on provides more than 30 tests covering all the nonparametric and categorical tests you normally use for larger datasets. Included are one-sample, two-sample, and K-Sample tests with independent or related samples, goodness-of-fit tests, tests of independence, and measures of association.

SPSS Categories

The SPSS Categories add-on is designed for you to reveal relationships among your categorical data. To help you understand your data, SPSS Categories uses perceptual mapping, optimal scaling, preference scaling, and dimension reduction. Using these techniques, you can visually interpret the relationships among your rows and columns.

SPSS Categories performs its analysis and displays results so you can understand ordinal and nominal data. It uses procedures similar to conventional regression, principal components, and canonical correlation. It performs regression using nominal or ordinal categorical predictor or outcome variables.

The procedures of SPSS Categories make it possible to perform statistical operations on categorical data:

- Using the scaling procedures, you can assign units of measurement and zero-points to your categorical data, which gives you access to new groups of statistical functions because you can analyze variables using mixed measurement levels.

- Using correspondence analysis, you can numerically evaluate similarities among nominal variables and summarize your data according to components you select.

- Using nonlinear canonical correlation analysis, you can collect variables of different measurement levels into sets of their own, and then analyze the sets.

This add-on can be used to produce a couple of very useful tools:

- ✓ **Perceptual maps** are high-resolution summary charts that serve as graphic displays of similar variables or categories. They give you insights into relationships among more than two categorical variables.

- ✓ **Biplots** are summary charts that make it possible to look at the relationships among products, customers, and demographic characteristics.

SPSS Conjoint

SPSS Conjoint provides a way for you to determine how each of your product's attributes affect consumer preference. When you combine conjoint analysis with competitive market product research, it's easier to zero in on product characteristics that are important to your customers.

With this research you can determine which product attributes your customers care about, which ones they care about most, and how you can do useful studies of pricing and brand equity. And you can do all this *before* incurring the expense of bringing new products to market.

SPSS Neural Networks

A *neural net* is a latticelike network of neuronlike nodes, set up within SPSS to act something like the neurons in a living brain. The connections between these nodes have associated *weights* (degrees of relative effect), which are adjustable. When you adjust the weight of a connection, the network is said to learn.

In the SPSS Neural Network, a training algorithm iteratively adjusts the weights to closely match the actual relationships among the data. The idea is to minimize errors and maximize accurate predictions. The computational neural network has one layer of neurons for input, another for output, with one or more hidden layers between them. The neural network is combined with other statistical procedures to provide clearer insight.

Using the familiar IBM SPSS Statistics interface, you can mine your data for relationships. After selecting a procedure, you specify the dependent variables, which may be any combination of scale and categorical types. To prepare for processing, you lay out the neural network architecture, including the computational resources you wish to apply. To complete preparation, you choose what to do with the output:

✔ List the results in tables

✔ Graphically display the results in charts

✔ Place the results in temporary variables in the dataset

✔ Export models in XML formatted files

SPSS Forecasting

You can use SPSS Forecasting to rapidly construct expert time-series forecasts. This module includes statistical algorithms you can use to analyze historical data and predict trends. You can set it up to analyze hundreds of different time series at once instead of running a separate procedure for each one.

The software is designed to handle the special situations that arise in trend analysis. It automatically determines the best-fitting ARIMA (Autoregressive Integrated Moving Average) or smoothing model. It automatically tests data for seasonality, intermittency, and missing values. The software detects outliers and prevents them from unduly influencing the results. The graphs generated include confidence intervals and indicate the model's goodness of fit.

As you gain experience at forecasting, SPSS Trends gives you more control over every parameter when you're building your data model. You can use the Expert Modeler in SPSS Forecasting to recommend starting points or to check calculations you've done by hand.

You can design models and save them in such a way that your forecasts can be updated on the arrival of changed data, or new data, without the necessity of re-estimating the model. Also, you can write scripts to update the models as situations change.

Chapter 20

Ten Useful SPSS Things You Can Find on the Internet

The names SPSS and PASW refer to the same software, so the names can be used interchangeably. The software now known as IBM SPSS Statistics was for a couple of years known as PASW, and before that was known for decades as SPSS. Because the name changes are recent, many of the resources you find on the Internet still use the name PASW.

The SPSS system is used in enough places and by enough people that it appears in many places on the Internet. Some of the Web pages are produced by the company that manufactures the software, but many pages are produced by people outside the company who are interested in using SPSS. This chapter gives you a general idea of the purpose of some of the most useful sites.

You may not want to type the URLs in this chapter, so I created a Web page that offers links you can click. Go to this book's associated Web site or to the following address:

```
www.belugalake.com/pasw
```

SPSS Humor

You will find an amazing variety of SPSS stuff on the Internet, from specific programming to general commentary. Even humor. The following two Web sites are dedicated to SPSS and statistics jokes:

```
www.ilstu.edu/%7egcramsey/Gallery.html
www.kingdouglas.com/SPSS/DiverseCultures/Humor.htm
```

The SPSS Home Page

The Web site of the SPSS company, from which you can find articles, programs, add-ons, and general news about SPSS, can be located by pointing your Web browser to the following address:

```
www.spss.com
```

Another way to get to the same Web site is to use the menus of SPSS and choose Help⇨SPSSInc Home.

From this base Web site, you can locate the SPSS home page for 34 countries other than the United States. This page allows you to specify a search string so you can locate the article, training service, or detailed description of whatever you want. The Web site is quite large and will probably contain some information about whatever it is you are trying to research, whether it's about the company SPSS, statistics in general, or SPSS in particular.

SPSS Developer Center

Whether you want to write SPSS programs or become otherwise knowledgeable about the workings of SPSS, you will need to check out the developer center. It has information on all sorts of SPSS operations. You can find the center here:

```
www.spss.com/devcentral/
```

You can use your Web browser to go directly to this site, or you can use the menu on the main window of SPSS and choose Help⇨Developer Central. Plenty of information is on that Web site, so you will need to browse around to find what you're looking for.

You can download utility programs already written and ready to go, graphics examples, new statistical modules, and articles on the inner workings of SPSS technology.

SPSS has forums where you can interact with people inside the SPSS company and with other SPSS users. If you have a question or a problem, this Web site is a good place to take it.

User Groups

SPSS has experts and experienced users, and a lot of them are ready to answer questions. If you have a question, don't just sit there with a giant question mark floating over your head, check out these sites:

```
www.spssusers.co.uk/
www.spsslog.com/
```

Mailing Lists and News Groups

A surprisingly large number of mailing lists are based on statistics. If you want, you can join a mailing list and receive copies of the ongoing discussions. You need not make your presence known until you have a question or have something to contribute. You can choose from among the mailing lists at the following sites:

- ✔ http://listserv.uark.edu/archives/ua-spss-user-group.html

- ✔ http://list.haifa.ac.il/mailman/listinfo/spss-users

- ✔ www.stattransfer.com/lists.html

The following is a newsgroup frequented by SPSS users:

```
comp.soft-sys.stat.spss
```

To take a look at examples of newsgroup postings, you can read the archived articles at the following location:

```
http://groups.google.com/group/comp.soft-sys.stat.spss
```

For statistics in general, three newsgroups exist. Following are the name and URL for the archived Web site of each one. You can look at the archives at these sites and get an idea of the type and frequency of posts:

- ✔ **sci.stat.consult:** `http://groups.google.com/group/sci.stat.consult`
- ✔ **sci.stat.edu:** `http://groups.google.com/group/sci.stat.edu`
- ✔ **sci.stat.math:** `http://groups.google.com/group/sci.stat.math`

Python Programming

This book gives you a small peek at the things you can do with Python. Although Python is a language built into SPSS, it's much more than that — even more than you will ever need for purely SPSS purposes. Python is a general-purpose programming language on the order of C or Java — that is, you can use it to do anything you might ever want to do with a computer.

And it runs almost anywhere. You will find versions of Python for Linux, Windows, Apple, and even cell phones. That's right. You can probably use it to program your cell phone.

If you want to go further into Python, there is no better place to start than the Python Language Web site. Lots of stuff is there, but two things are of prime importance: complete documentation (tutorials, examples, and more) and a free copy of Python that you can download and install on your machine:

```
www.python.org
```

Quite often, newcomers to programming find themselves put off by the geeky terms used to describe a programming language. Don't be. It's a lot easier to understand the fundamentals of programming than it is to understand statistics; it's just that nerds like to show off by talking that way. (I should know. I've spent my life in the company of nerds.)

The Python Web site is friendlier than most of its kind; it's an excellent place to start learning programming. Programming is a pretty good hobby, but it can be habit forming. Be careful — you can find yourself getting hooked, and before you know it, you're on the road to becoming a nerd.

The following Web sites are helpful when you're programming Python within SPSS:

- ✔ www.python.org
- ✔ www.spsstools.net/python.htm
- ✔ www.spss.com/devcentral
- ✔ http://training.spss.co.uk/pdf/Introduction_to_PASW_Statistics_Programmability_Python_Programs.pdf

Script and Syntax Programming

You can find programs and programming tutorials for the various SPSS languages. All the Web sites listed here concern themselves with programming SPSS. Most have commentary and suggestions along with programs, some are tutorials on programming, and some have programs that you can download and use.

Syntax language:

- ✔ www.ats.ucla.edu/stat/spss/seminars/spss_syntax/
- ✔ http://bama.ua.edu/~jhartman/689/syntax.ppt
- ✔ www.hmdc.harvard.edu/pub_files/SPSS_Syntax.pdf"
- ✔ www.longitudinal.stir.ac.uk/SPSS_support.html
- ✔ www.spsstools.net/

Scripts (Sax BASIC):

- ✔ http://pages.infinit.net/rlevesqu/SampleScripts.htm
- ✔ http://ftgsoftware.com/manuals/basic32.pdf
- ✔ www.ocair.org/files/VBawksp/spss.htm
- ✔ www.spssusers.co.uk/Tips/saxbasic_doc.html

General SPSS programming:

- ✔ www.spsstools.net/
- ✔ www.spss.com/downloads/Papers.cfm?List=all&Name=all
- ✔ http://scripts.filehungry.com/product/java/javabeans/development_tools/java_spss_writer

Tutorials for SPSS and Statistics

One of the things the Web does very well is present tutorials. In fact, that's the sort of thing it was originally designed to do — instead of the advertising and marketing tool that it has become. This section contains a short list of tutorial Web sites, but there are certainly more. Some of the sites are for statistics, some are for SPSS, and some are for both.

If you're looking for a tutorial, you'll probably need to search through several of these sites to find the one you want to start with. Some are better than others. They all emphasize certain characteristics and capabilities of the software. Some specialize in statistics for a particular subject, which may or may not be to your advantage. Some were designed using older versions of the software, but the capabilities of SPSS have expanded, not contracted, so those lessons should still be valid.

This list is only a small percentage of the total. These are mostly for general-purpose studies, but some sites become specific in the types of statistics they present. If you wanted to narrow your search to say, medical statistics, you could enter the search string *SPSS tutorial medical* or *PASW tutorial medical* to turn up a number of specialized sites.

SPSS tutorials:

- www.hmdc.harvard.edu/projects/SPSS_Tutorial/spsstut.shtml
- http://calcnet.mth.cmich.edu/org/spss/toc.htm
- www.utexas.edu/its/rc/tutorials/stat/spss/spss1/
- www.shef.ac.uk/scharr/spss/
- www.statisticsolutions.com/SPSS-tutorial
- www.students.stir.ac.uk/docs/spss/spss.html
- www.datastep.com/SPSSTraining.html
- www.stat.tamu.edu/spss.php
- http://academic.uofs.edu/department/psych/methods/cannon99/spssmain.html
- http://its.unm.edu/introductions/Spss_tutorial/
- www.uni.edu/its/us/document/stats/spss2.html

General statistics tutorials:

- ✔ www2.chass.ncsu.edu/garson/pa765/statnote.htm
- ✔ www.meandeviation.com
- ✔ http://davidmlane.com/hyperstat/
- ✔ www.psych.utoronto.ca/courses/c1/statstoc.htm
- ✔ www.statsoft.com/textbook/stathome.html
- ✔ http://mail.pittstate.edu/~winters/tutorial/
- ✔ http://math.about.com/od/statistics/Statistics_
  Tutorials_and_Resources.htm
- ✔ www.tulane.edu/~panda2/Analysis2/ahome.html

SPSS Wiki

A *wiki* is a Web site with documents that are constantly updated. You can join as a reader and as a contributor. The SPSS wiki acts both as a reference source and as a workbook for SPSS statistical procedures. It can be used equally well by both novices and experts.

Instructions on the Web page tell you how to use the wiki to find what you're looking for — and how to contribute to the constantly growing body of information. You will find the SPSS wiki at the following location:

```
http://spss.wikia.com
```

PSPP, a Free SPSS

You have probably heard of the Free Software Foundation and GNU. The members are involved in developing open-source software (to me and you, that translates into both "free of charge" and "free to modify"). The PSPP project is developing a SPSS workalike. It's not possible for me to say how much has been finished and tested, because that changes almost daily, but claims are being made that it supports a large subset of SPSS. Its statistical procedure support is limited but growing.

I'm not recommending it, but I'm not pooh-poohing it either. If you're interested, you can download a copy and try it for yourself. It can be downloaded in different ways and installed in different forms. You can find out all about how to do that at this Web site:

```
www.gnu.org/software/pspp/
```

You can get the latest stable version or you can get a copy of the current version while it's under development. I recommend that you get the latest stable version, at least to begin with, unless you are either a programmer or love surprises.

Besides the normal descriptive text found on the Web site, you will find e-mail addresses and IRC channels for discussions and support. You can register to be notified of future releases.

Glossary

Add-on: A utility that can be added to SPSS. Also called a *module*.

Analysis of covariance: See *ANCOVA*.

Analysis of variance: See *ANOVA*.

ANCOVA: Analysis of covariance. ANOVA with the addition of a second or third covariate.

ANOVA: Analysis of variance. Using an F-ratio to test the fit of a linear model.

ascending: A sorting order. The cases are ordered so the values range from small to large. See also *descending*.

association: Variables are said to be *associated* if the value of one is a whole or fractional multiplication of the other.

autoscript: A script that executes automatically in response to the output of data. Specific autoscripts can be assigned to specific output types. See also *script*.

average: The result of adding several values and then dividing by the number of values. See also *mean* and *mode*.

base: The main system of SPSS. Modules can be added to expand SPSS, but the base system is always present.

BASIC: See *script*.

bell curve: See *normal distribution*.

binning: The process of organizing the values of a variable into groups. Each group is a defined as a specific range of values and each group can be thought of as being sorted a bin. This is also called *clustering*.

bivariate: Using two variables.

break variable: When organizing data into tabular form, the break variable is used to group the information. At the point in the report where the break variable changes value, a subtotal line is generated, or a new page is started, or some other break appears in the report.

canonical correlation: A correlation expressed in a standard form.

case: Any single collection of values. All the values in a single row. A case is sometimes called a single record, and it normally contains one constant value for each variable.

case summary: A simple table that directly summarizes values of the cases.

categorical variable: A type of variable that can take on only one of a specific set of values, such as year of birth, make of car, or favorite color — in effect, defining a category. See also *scale, ordinal, nominal, dichotomy,* and *binning.*

censored case: A case for which the event being analyzed has not occurred during the time period of the study.

chart: See *graph.*

clustering: See *binning.*

coefficient of determination: A statistic used to specify the correctness of the fit of regression coefficients.

command language: See *Syntax.*

confidence interval: A range above and below an average into which a specified percentage of the values appears. For example, if gravel trucks for a company deliver an average of 190 loads per month, but 95 percent of the trucks deliver between 183 and 194 loads, the 95 percent confidence interval ranges from a low of 7 below to a high of 4 above.

constant: A number. A quantity that is regarded as fixed or unchanging. See also *variable.*

continuous: See *scale.*

correlation: The degree of similarity or difference between two variables.

covariance: A comparison of the variance of one set of values with that of another.

covariate: A variable that takes part in the prediction of an outcome. An *independent variable* in *regression.* It is secondary to the relationship of the main independent variable.

cutpoint: A number used as a divider to split values into groups, as in *binning.*

dataset: The data displayed in the Data Editor window, whether loaded from a file, entered from the keyboard, or both. (It's also written as two words: data set.) Multiple datasets can be loaded and will appear in separate windows — they will be labeled DataSet1, DataSet2, and so on.

degrees of freedom: The minimum number of values that must be specified to determine all the data points. This number is usually one less than the number of values used in the calculation.

delimiter: A character used to indicate the beginning of, ending of, or separation between individual values in a series of strings of characters. For example, the string of characters 59,21,34 is a series of comma-delimited numbers.

dependent variable: A variable that has its value derived from one or more other variables. Also called a *predicted variable*. See also *independent variable*.

descending: A sorting order that arranges values from large to small. See also *ascending.*

deviation: The amount by which a measurement differs from some fixed value.

dichotomy: A variable with only two possible values, such as yes/no, true/false, or like/dislike. It is a specific type of categorical variable. See also *categorical variable.*

dodging: Plotting points on a graph so they appear next to one another instead of one of top of the other.

error: Two kinds of errors exist in the world of statistics. The conventional kind comes about when you enter a wrong number and get a bogus result. The other kind is calculated — that is, you calculate the amount of possible error present in the results you get from the data you have. With modern survey techniques, you will often hear the term "margin of error" for this second type.

faceting: See *paneling.*

field: In the SPSS documentation, *field* is used as a synonym for *variable.*

F-ratio: A comparison of the variance of unexpected values to the variance of expected values.

frequency distribution: The collection of values that a variable takes in a sample.

GLM: General Linear Model. A general procedure for analyzing *variance*, *covariance*, and *regression*.

goodness of fit: The extent to which observed values approximate values from a theoretical distribution.

graph: A non-numeric display of values. The terms *graph* and *chart* are used in SPSS internal documentation almost interchangeably.

GUI: Graphical User Interface. Control of an application with windows and a mouse. All versions of SPSS operate this way.

histogram: A graphical display of a distribution in which the extent of each rectangle represents the magnitude (as in a bar chart) and the width of each rectangle represents the magnitude of the bin. The area of each rectangle thus represents the frequency.

hoc: See *post hoc*.

imputation: The process of calculating numeric values for missing values in the data.

independence: The degree to which two or more variables have no effect on one another.

independent variable: A variable whose values are used as the basis for calculation of statistics. See also *dependent variable*.

kurtosis: A measure of how peaked a bell curve is. A positive number indicates there is more of a peak than standard; a negative number indicates a flatter line.

Levene test: A test that determines whether the *variance* of two groups is significantly different or significantly the same.

linear: A straight line. No curves.

log-linear model: An analysis based on a correlation using the raw values of one variable and the natural logarithm of another.

longitudinal data: Data which spans all cases. Not clustered.

mean: 1. Another word for average. 2. A calculated value equally distant from the two extreme values. 3. The temperament of the person making you learn this stuff. See also *average* and *mode*.

missing data: If you declare a value for a variable as representing the fact that no value is present, the missing value will not be included in calculations.

mixed model: A statistical model containing both fixed effects and random effects.

mode: The value that occurs most frequently in a given set of data. See also *average* and *mean*.

model: a mathematical model of some process.

module: A utility that can be added to SPSS. Also called an *add-on*.

multiple response set: A special variable that has its content generated from the content of two or more other variables. In SPSS, it doesn't appear in the Data View (in the Data Editor window), but does appear when you select variable names for other activities.

multivariate: Using multiple variables.

nominal: Numbers that specify categories. For example, yes, no, and undecided could be represented by 2, 1, and 0. See also *scale, ordinal,* and *categorical*.

nonlinear: Not in a straight line. Curved.

normality: The degree to which the values match *normal distribution*.

normal distribution: A distribution that is continuous and symmetric. It is used primarily because many quantitative measurements appear to approximate this distribution. It is also called the bell curve.

OLAP cubes: Online Analytical Processing cubes. A multilevel table containing totals, *means*, or some other statistics in which each level of the table contains the values relating to one value of a *categorical variable*.

OMS: Output Management Systems. The ability in SPSS to output to different file formats.

Online Analytical Processing: See *OLAP cubes*.

ordinal: Types of numbers that specify the order of occurrences. The ordinal forms of 1, 2, and 3 are first, second, and third. See also *scale, nominal,* and *categorical*.

outliers: The extreme values of a variable. Generally, they are the five largest and five smallest values.

paneling: Adding another dimension of data to a graphic display causing the layout to be replicated a number of times to accommodate the values of the data along the new dimension. This process is also known as *faceting*.

parametric: A procedure that requires one or more seed values that control its processing.

PASW: Predictive Analysis SoftWare. For a couple of years, SPSS was known as PASW.

Pearson's Product Moment Correlation: Commonly called Pearson's correlation. It represents the degree of linear relationship between two variables.

periodicity: The interval of repetition at which data recordings are made.

pivot table: A table with names identifying the rows and columns. Swapping the positions of the rows and columns to make the table appear in a different form, but containing the same data, is known as *pivoting* the table. The tables in SPSS Viewer are pivot tables.

post hoc: The erroneous conclusion that some condition arises as the result of a previous condition.

p-p plot: A proportion-proportion plot. The observed cumulative proportion is plotted against the expected cumulative proportion.

predicted variable: See *dependent variable*.

predictor: A variable, or collection of variables, the values of which predict the values of some others.

probit: A function of probability based on the quartiles of normal distribution.

pyramid: A special form of a histogram where the bars representing the values extend outward to the sides from a center line. It often assumes the shape of a pyramid.

Python: A general-purpose programming language that can also be used to program SPSS scripting.

q-q plot: A quantile-quantile plot. The quantiles of the observed values are plotted against the quantiles of a specified distribution.

quantiles: A set of values chosen to divide a sampling of data into groups, each containing (as far as possible) an equal number of values.

quartile: Specific values that divide all the values into four groups, with an equal number of values in each group. The groups are generally called the first, second, third, and fourth quartiles.

R: See *coefficient of determination.*

recency: The quality or state of being recent.

recoding: The conversion of a set of SPSS values to a new set of values. For example, if you have yes/no coded as 0/1, by recoding you can change the values to 1/2 in a single operation.

record: Any single collection of values for the variables defined in SPSS. A record is all the values of a single row. It is a single case or row.

regression: Determining the "best fit" equation for the relationship between two variables. See also *dependent variable* and *independent variable.*

row: Any single collection of values for all the variables defined in a SPSS dataset. It appears as a single row in the Data View window. It is a single case.

scale: A type of number that uses a standard by which something is measured, such as inches, pounds, dollars, or hours. Another name for scale is *continuous.* See also *ordinal, nominal,* and *categorical.*

script: A program written in either the BASIC or Python programming language. These are different languages than Syntax.

skewness: A measure of the unevenness of the distribution of data. Positive skewness indicates more high values than low in the distribution; negative skewness indicates more low values than high.

SPSS: Statistical Package for the Social Sciences. The original name of SPSS.

standard deviation: A calculated indicator of the extent of deviation for a specific collection of data. The value is derived from the variations where the points are compared to a standard bell-shaped curve. It is the square root of the *variance.*

standard error: A measurement of the magnitude of the change from one sample to the next.

statistic: A single number calculated in a specific way. Some examples of types of a statistics are sum, *mean, deviation,* and *average.*

statistics: A collection of statistical values.

string: A series of characters making up a name or even a complete sentence. Quite often the beginning and ending of a string is delimited by quotes.

Syntax: The name of the programming language fundamental to SPSS. All actions performed by SPSS are in response to the internal interpretation of Syntax commands. In the SPSS documentation, Syntax is sometimes referred to as the *command language*.

t: The number of degrees of freedom. A continuous distribution with density symmetrical around the null value and a bell-shaped curve.

univariate: A statistic derived from the values of one variable. Examples are *mean*, *standard deviation*, and sum.

variable: In statistical software, a place to store constants. A variable can store a number of constants (one for each case). Each case (or row) in SPSS consists of a collection of constant values assigned to variables.

variance: The average of the differences between a set of measured values and a set of expected values on a standard bell-shaped curve. It is the square of the *standard deviation.*

Index

Business/Accounting & Bookkeeping

Bookkeeping For Dummies
978-0-7645-9848-7

eBay Business
All-in-One For Dummies,
2nd Edition
978-0-470-38536-4

Job Interviews
For Dummies,
3rd Edition
978-0-470-17748-8

Resumes For Dummies,
5th Edition
978-0-470-08037-5

Stock Investing
For Dummies,
3rd Edition
978-0-470-40114-9

Successful Time
Management
For Dummies
978-0-470-29034-7

Computer Hardware

BlackBerry For Dummies,
3rd Edition
978-0-470-45762-7

Computers For Seniors
For Dummies
978-0-470-24055-7

iPhone For Dummies,
2nd Edition
978-0-470-42342-4

Laptops For Dummies,
3rd Edition
978-0-470-27759-1

Macs For Dummies,
10th Edition
978-0-470-27817-8

Cooking & Entertaining

Cooking Basics
For Dummies,
3rd Edition
978-0-7645-7206-7

Wine For Dummies,
4th Edition
978-0-470-04579-4

Diet & Nutrition

Dieting For Dummies,
2nd Edition
978-0-7645-4149-0

Nutrition For Dummies,
4th Edition
978-0-471-79868-2

Weight Training
For Dummies,
3rd Edition
978-0-471-76845-6

Digital Photography

Digital Photography
For Dummies,
6th Edition
978-0-470-25074-7

Photoshop Elements 7
For Dummies
978-0-470-39700-8

Gardening

Gardening Basics
For Dummies
978-0-470-03749-2

Organic Gardening
For Dummies,
2nd Edition
978-0-470-43067-5

Green/Sustainable

Green Building
& Remodeling
For Dummies
978-0-470-17559-0

Green Cleaning
For Dummies
978-0-470-39106-8

Green IT For Dummies
978-0-470-38688-0

Health

Diabetes For Dummies,
3rd Edition
978-0-470-27086-8

Food Allergies
For Dummies
978-0-470-09584-3

Living Gluten-Free
For Dummies
978-0-471-77383-2

Hobbies/General

Chess For Dummies,
2nd Edition
978-0-7645-8404-6

Drawing For Dummies
978-0-7645-5476-6

Knitting For Dummies,
2nd Edition
978-0-470-28747-7

Organizing For Dummies
978-0-7645-5300-4

SuDoku For Dummies
978-0-470-01892-7

Home Improvement

Energy Efficient Homes
For Dummies
978-0-470-37602-7

Home Theater
For Dummies,
3rd Edition
978-0-470-41189-6

Living the Country Lifestyle
All-in-One For Dummies
978-0-470-43061-3

Solar Power Your Home
For Dummies
978-0-470-17569-9

Internet

Blogging For Dummies,
2nd Edition
978-0-470-23017-6

eBay For Dummies,
6th Edition
978-0-470-49741-8

Facebook For Dummies
978-0-470-26273-3

Google Blogger
For Dummies
978-0-470-40742-4

Web Marketing
For Dummies,
2nd Edition
978-0-470-37181-7

WordPress For Dummies,
2nd Edition
978-0-470-40296-2

Language & Foreign Language

French For Dummies
978-0-7645-5193-2

Italian Phrases
For Dummies
978-0-7645-7203-6

Spanish For Dummies
978-0-7645-5194-9

Spanish For Dummies,
Audio Set
978-0-470-09585-0

Macintosh

Mac OS X Snow Leopard
For Dummies
978-0-470-43543-4

Math & Science

Algebra I For Dummies
978-0-7645-5325-7

Biology For Dummies
978-0-7645-5326-4

Calculus For Dummies
978-0-7645-2498-1

Chemistry For Dummies
978-0-7645-5430-8

Microsoft Office

Excel 2007 For Dummies
978-0-470-03737-9

Office 2007 All-in-One
Desk Reference
For Dummies
978-0-471-78279-7

Music

Guitar For Dummies,
2nd Edition
978-0-7645-9904-0

iPod & iTunes
For Dummies,
6th Edition
978-0-470-39062-7

Piano Exercises
For Dummies
978-0-470-38765-8

Parenting & Education

Parenting For Dummies,
2nd Edition
978-0-7645-5418-6

Type 1 Diabetes
For Dummies
978-0-470-17811-9

Pets

Cats For Dummies,
2nd Edition
978-0-7645-5275-5

Dog Training For Dummies,
2nd Edition
978-0-7645-8418-3

Puppies For Dummies,
2nd Edition
978-0-470-03717-1

Religion & Inspiration

The Bible For Dummies
978-0-7645-5296-0

Catholicism For Dummies
978-0-7645-5391-2

Women in the Bible
For Dummies
978-0-7645-8475-6

Self-Help & Relationship

Anger Management
For Dummies
978-0-470-03715-7

Overcoming Anxiety
For Dummies
978-0-7645-5447-6

Sports

Baseball For Dummies,
3rd Edition
978-0-7645-7537-2

Basketball For Dummies,
2nd Edition
978-0-7645-5248-9

Golf For Dummies,
3rd Edition
978-0-471-76871-5

Web Development

Web Design All-in-One
For Dummies
978-0-470-41796-6

Windows Vista

Windows Vista
For Dummies
978-0-471-75421-3

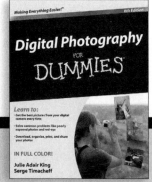

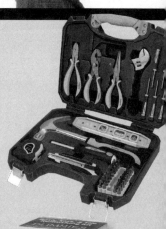